Control and Analysis for Food and Agricultural Products

Control and Analysis for Food and Agricultural Products

Editor
Kavita Marwaha

2010
Gene-Tech Books
New Delhi - 110 002

ISBN 81-89729-93-4
ISBN 978-81-89729-93-6

Published by : **GENE-TECH BOOKS**
4762-63/23, Ansari Road, Darya Ganj,
NEW DELHI - 110 002
Phone: 41562849
e-mail: genetechbooks@yahoo.co.in

Printed at : **Chawla Offset Printers**
New Delhi – 110 052

PRINTED IN INDIA

Preface

Agriculture contemporarily involves the use of the latest scientific principles and methodology, constantly improvising upon itself to meet the ever-growing demand for food and fuel. In such a scenario, analysis and control methods for farm practices and production has been a considerable increase in utilisation, particularly through the emergence of such sciences and precision agriculture, which involves the use of state-of-the-art technology in field practice agricultural needs as per spatial variability of the farm. As such, analysis and control methods for farm management entails the next level for agriculture, heightening its technological revolution.

This book has been designed keeping in mind the need of a text which acquaint readers with the emergence of analysis and control procedures in agricultural practice. The book highlights the principles and techniques which define this science, bringing within purview the changes, advantages and challenges offered by it. A comprehensive discussion of sampling methods, measurement devices, sensors, computer analysis, data interpretation standardisation, etc., has been brought under cover, along with a review of their efficacy in agriculture. Upto-date with the latest in the field, the text manages to incorporate a pertinent account of what analysis and control methods in agriculture offers for the present as well as the future.

Kavita Marwaha

Contents

	Preface	*v*
1.	Principles of Food Analysis	1
2.	Quality Control in Food Industry	17
3.	Food Sampling and Data Analysis	51
4.	Food Protein Analysis	69
5.	Analysis of Lipid Concentration in Food	89
6.	Determination of Ash and Minerals	115
7.	Analysis of Moisture and Total Solids in Food Products	133
8.	Instrumental Techniques for Food Analysis	151
9.	Evaluation of Milk Quality	189
10.	Quality Management in Fish Processing	215
11.	Microbial Safety in Food Preservation	233
12.	Food Safety Risk Analysis	243
	Bibliography	*269*
	Index	*271*

1

Principles of Food Analysis

Food analysis is the discipline dealing with the development, application and study of analytical procedures for characterising the properties of foods and their constituents. These analytical procedures are used to provide information about a wide variety of different characteristics of foods, including their composition, structure, physicochemical properties and sensory attributes. This information is critical to our rational understanding of the factors that determine the properties of foods, as well as to our ability to economically produce foods that are consistently safe, nutritious and desirable and for consumers to make informed choices about their diet.

Foods are analysed by scientists working in all of the major sectors of the food industry including food manufacturers, ingredient suppliers, analytical service laboratories, government laboratories, and University research laboratories.

Government Regulations and Recommendations

Government regulations and recommendations are designed to maintain the general quality of the food supply, to ensure the food industry provides consumers with foods that are wholesome and safe, to inform consumers about the nutritional composition of foods so that they can make knowledgeable choices about their diet, to enable fair competition amongst food companies, and to eliminate economic fraud.

There are a number of Government Departments Responsible for regulating the composition and quality of foods, including the Food and Drug Administration (FDA), the United States Department of Agriculture (USDA), the National Marine Fisheries Service (NMFS) and the Environmental Protection Agency (EPA). Each of these government agencies is responsible for regulating particular sectors of the food industry and publishes documents that contain detailed information about the regulations and recommendations pertaining to the foods produced within those sectors.

Voluntary and Mandatory Standards

Government agencies have specified a number of voluntary and mandatory standards concerning the composition, quality, inspection, and labelling of specific food products.

Mandatory Standards:

— *Standards of Identity:* These regulations specify the type and amounts of ingredients that certain foods must contain if they are to be called by a particular name on the food label. For some foods there is a maximum or minimum concentration of a certain component that they must contain, e.g., "peanut butter" must be less than 55% fat, "ice-cream" must be greater than 10% milk fat, "cheddar cheese" must be greater than 50% milk fat and less than 39% moisture.

— *Standards of Quality*: Standards of quality have been defined for certain foods (e.g., canned fruits and vegetables) to set minimum requirements on the colour, tenderness, mass and freedom from defects.

— *Standards of Fill-of-Container:* These standards state how full a container must be to avoid consumer deception, as well as specifying how the degree of fill is measured.

Voluntary Standards:

— *Standards of Grade:* A number of foods, including meat, dairy products and eggs, are graded according to their quality, e.g. from standard to excellent. For example meats can be graded as "prime", "choice", "select", "standard" etc according to their origin, tenderness, juiciness, flavour and appearance. There are clear definitions associated with these descriptors that products must conform to before they can be given the appropriate label. Specification of the grade of a food product on the label is voluntary, but many food manufacturers opt to do this because superior grade products can be sold for a higher price. The government has laboratories that food producers send their products too to be tested to receive the appropriate certification. This service is requested and paid for by the food producer.

Nutritional Labelling and Education Act

In 1990, the US government passed the Nutritional Labelling and Education Act (NLEA), which revised the regulations pertaining to the nutritional labelling of foods, and made it mandatory for almost all food products to have standardised nutritional labels. One of the major reasons for introducing these regulations was so that consumers could make informed choices about their diet. Nutritional labels state the total calorific value of the food, as well as total fat, saturated fat, cholesterol, sodium, carbohydrate, dietary fibber, sugars, protein, vitamins, calcium and iron.

The label may also contain information about nutrient content claims (such as "low fat", "low sodium" "high fibber" "fat free" etc), although government regulations stipulate the minimum or maximum amounts of specific food components that a food must contain if it is to be given one of these nutrient content descriptors. The label may also contain certain FDA approved health claims based on links between specific food components and certain diseases. The information provided on the label can be used by consumers to plan a nutritious and balanced diet, to avoid over consumption of food components linked with health problems, and to encourage greater consumption of foods that are beneficial to health.

Authenticity

The price of certain foods is dictated by the quality of the ingredients that they contain. For example, a packet of premium coffee may claim that the coffee beans are from Columbia, or the label of an expensive wine may claim that it was produced in a certain region, using a certain type of grapes in a particular year. There are many instances in the past where manufacturers have made false claims about the authenticity of their products in order to get a higher price. It is therefore important to have analytical techniques that can be used to test the authenticity of certain food components, to ensure that consumers are not the victims of economic fraud and that competition among food manufacturers is fair.

Food Inspection and Grading Services

The government has a Food Inspection and Grading Service that routinely analyses the properties of food products to ensure that they meet the appropriate laws and regulations. Hence, both government agencies and food manufacturers need analytical techniques to provide the appropriate information about food properties. The most important criteria for this type of test are often the accuracy of the measurements and the use of an official method.

The government has recently carried out a survey of many of the official analytical techniques developed to analyse foods, and has specified which techniques must be used to analyse certain food components for labelling purposes. Techniques have been chosen which provide accurate and reliable results, but which are relatively simple and inexpensive to perform.

Analysing Foods

One of the most important reasons for analysing foods from both the consumers and the manufacturers standpoint is to ensure that they are safe. It would be economically disastrous, as well as being rather unpleasant to consumers, if a food manufacturer sold a product that was harmful or toxic. A food may be considered to be unsafe because it contains harmful microorganisms, toxic chemicals or extraneous matter. It is therefore important that food manufacturers do everything they can to ensure that these harmful substances are not present, or that they are effectively eliminated before the food is consumed.

This can be achieved by following "good manufacturing practice" regulations specified by the government for specific food products and by having analytical techniques that are capable of detecting harmful substances. In many situations it is important to use analytical techniques that have a high sensitivity, i.e., that can reliably detect low levels of harmful material. Food manufacturers and government laboratories routinely analyse food products to ensure that they do not contain harmful substances and that the food production facility is operating correctly.

Quality Control

The food industry is highly competitive and food manufacturers are continually trying to increase their market-share and profits. To do this they must ensure that their products are of higher quality, less expensive, and more desirable than their competitors, whilst ensuring that they are safe and nutritious. To meet these rigorous standards food manufacturers need analytical techniques to analyse food materials before, during and after the manufacturing process to ensure that the final product meets the desired standards. In a food factory one starts with a number of different raw materials, processes them in a certain manner (e.g. heat, cool, mix, dry), packages them for consumption and then stores them. The food is then transported to a warehouse or retailer where it is sold for consumption.

One of the most important concerns of the food manufacturer is to produce a final product that consistently has the same overall properties, i.e. appearance, texture, flavour and shelf life. When we purchase a particular food product we expect its properties to be the same (or very similar) to previous times, and not to vary from purchase-to-purchase. Ideally, a food manufacture wants to take the raw ingredients, process them in a certain way and produce a product with specific desirable properties. Unfortunately, the properties of the raw ingredients and the processing conditions vary from time to time which causes the properties of the final product to vary, often in an unpredictable way.

How can food manufacturers control these variations? Firstly, they can understand the role that different food ingredients and processing operations play in determining the final properties of foods, so that they can rationally control the manufacturing process to produce a final product with consistent properties. This type of information can be established through research and development work. Secondly, they can monitor the properties of foods during production to ensure that they are meeting the specified requirements, and if a problem is detected during the production process, appropriate actions can be taken to maintain final product quality.

Characterisation of raw materials: Manufacturers measure the properties of incoming raw materials to ensure that they meet certain minimum standards of quality that have previously been defined by the manufacturer. If these standards are not met the manufacturer rejects the material. Even when a batch of raw materials has been accepted, variations in its properties might lead to changes in the properties of the final product. By analysing the raw materials it is often possible to predict their subsequent behaviour during processing so that the processing conditions can be altered to produce a final product with the desired properties. For example, the colour of potato chips depends on the concentration of reducing sugars in the potatoes that they are manufactured from: the higher the concentration, the browner the potato chip. Thus it is necessary to have an analytical technique to measure the concentration of reducing sugars in the potatoes so that the frying conditions can be altered to produce the optimum coloured potato chip.

Monitoring of food properties during processing: It is advantageous for food manufacturers to be able to measure the properties of foods during processing. Thus, if any problem develops, then it can be quickly detected, and the process adjusted to compensate for it. This helps to improve the overall quality of a food and to reduce the amount of material and time wasted. For example, if a manufacturer were producing a salad dressing product, and the oil content became too high or too low they would want to adjust the processing conditions to eliminate this problem. Traditionally, samples are removed from the process and tested in a quality assurance laboratory.

This procedure is often fairly time-consuming and means that some of the product is usually wasted before a particular problem becomes apparent. For this reason, there is an increasing tendency in the food industry to use analytical techniques which are capable of rapidly measuring the properties of foods on-line, without having to remove a sample from the process. These techniques allow problems to be determined much more quickly and therefore lead to improved product quality and less waste. The ideal criteria for an on-line technique is that it be capable of rapid and precise measurements, it is non-intrusive, it is nondestructive and that it can be automated.

Characterisation of final product: Once the product has been made it is important to analyse its properties to ensure that it meets the appropriate legal and labelling requirements, that it is safe, and that it is of high quality. It is also important to ensure that it retains its desirable properties up to the time when it is consumed. A system known as Hazard Analysis and Critical Control Point (HACCP) has been developed, whose aim is to systematically identify the ingredients or processes that may cause problems (hazard analysis), assign locations (critical control points) within the manufacturing process where the properties of the food must be measured to ensure that safety and quality are maintained, and to specify the appropriate action to take if a problem is identified.

The type of analytical technique required to carry out the analysis is often specified.

Properties Analysed

Food analysts are interested in obtaining information about a variety of different characteristics of foods, including their composition, structure, physicochemical properties and sensory attributes.

Composition of Food

The composition of a food largely determines its safety, nutrition, physicochemical properties, quality attributes and sensory characteristics. Most foods are compositionally complex materials made up of a wide variety of different chemical constituents. Their composition can be specified in a number of different ways depending on the property that is of interest to the analyst and the type of analytical procedure used: specific atoms; specific molecules, types of molecules, or specific substances. Government regulations state that the concentration of certain food components must be stipulated on the nutritional label of most food products, and are usually reported as specific molecules or types of molecules.

The structural organisation of the components within a food also plays a large role in determining the physicochemical properties, quality attributes and sensory characteristics of many foods. Hence, two foods that have the same composition can have very different quality attributes if their constituents are organised differently. For example, a carton of ice cream taken from a refrigerator has a pleasant appearance and good taste, but if it is allowed to melt and then is placed back in the refrigerator its appearance and texture change dramatically and it would not be acceptable to a consumer.

Thus, there has been an adverse influence on its quality, even though its chemical composition is unchanged, because of an alteration in the structural organisation of the constituents caused by the melting of ice and fat crystals.

Another familiar example is the change in egg white from a transparent viscous liquid to an optically opaque gel when it is heated in boiling water for a few minutes. Again there is no change in the chemical composition of the food, but its physiochemical properties have changed dramatically because of an alteration in the structural organisation of the constituents caused by protein unfolding and gelation.

The structure of a food can be examined at a number of different levels:

— Molecular structure (1 - 100 nm). Ultimately, the overall physicochemical properties of a food depend on the type of molecules present, their three-dimensional structure and their interactions with each other. It is therefore important for food scientists to have analytical techniques to examine the structure and interactions of individual food molecules.

— Microscopic structure (10 nm - 100 m). The microscopic structure of a food can be observed by microscopy (but not by the unaided eye) and consists of regions in a material where the molecules associate to form discrete phases, e.g., emulsion droplets, fat crystals, protein aggregates and small air cells.

— Macroscopic structure (> 100 m). This is the structure that can be observed by the unaided human eye, e.g., sugar granules, large air cells, raisons, chocolate chips

All of these different levels of structure contribute to the overall properties of foods, such as texture, appearance, stability and taste. In order to design new foods, or to improve the properties of existing foods, it is extremely useful to understand the relationship between the structural properties of foods and their bulk properties. Analytical techniques are therefore needed to characterise these different levels of structure. A number of the most important of these techniques are considered in this course.

Physicochemical Properties of Foods

The physiochemical properties of foods (rheological, optical, stability, "flavour") ultimately determine their perceived quality, sensory attributes and behaviour during production, storage and consumption.

- The optical properties of foods are determined by the way that they interact with electromagnetic radiation in the visible region of the spectrum, e.g., absorption, scattering, transmission and reflection of light. For example, full fat milk has a "whiter" appearance than skim milk because a greater fraction of the light incident upon the surface of full fat milk is scattered due to the presence of the fat droplets.
- The rheological properties of foods are determined by the way that the shape of the food changes, or the way that the food flows, in response to some applied force. For example, margarine should be spreadable when it comes out of a refrigerator, but it must not be so soft that it collapses under its own weight when it is left on a table.
- The stability of a food is a measure of its ability to resist changes in its properties over time. These changes may be chemical, physical or biological in origin. Chemical stability refers to the change in the type of molecules present in a food with time due to chemical or biochemical reactions, e.g., fat rancidity or non-enzymatic browning. Physical stability refers to the change in the spatial distribution of the molecules present in a food with time due to movement of molecules from one location to another, e.g., droplet creaming in milk. Biological stability refers to the change in the number of microorganisms present in a food with time, e.g., bacterial or fungal growth.
- The flavour of a food is determined by the way that certain molecules in the food interact with receptors in the mouth (taste) and nose (smell) of human beings. The perceived flavour of a food product depends on the type and concentration of flavour constituents within it, the nature of the food matrix, as well as how quickly the flavour molecules can move from the food to the

sensors in the mouth and nose. Analytically, the flavour of a food is often characterised by measuring the concentration, type and release of flavour molecules within a food or in the headspace above the food.

Foods must therefore be carefully designed so that they have the required physicochemical properties over the range of environmental conditions that they will experience during processing, storage and consumption, e.g., variations in temperature or mechanical stress. Consequently, analytical techniques are needed to test foods to ensure that they have the appropriate physicochemical properties.

Sensory Attributes

Ultimately, the quality and desirability of a food product is determined by its interaction with the sensory organs of human beings, e.g., vision, taste, smell, feel and hearing. For this reason the sensory properties of new or improved foods are usually tested by human beings to ensure that they have acceptable and desirable properties before they are launched onto the market. Even so, individuals' perceptions of sensory attributes are often fairly subjective, being influenced by such factors as current trends, nutritional education, climate, age, health, and social, cultural and religious patterns.

To minimise the effects of such factors a number of procedures have been developed to obtain statistically relevant information. For example, foods are often tested on statistically large groups of untrained consumers to determine their reaction to a new or improved product before full-scale marketing or further development. Alternatively, selected individuals may be trained so that they can reliably detect small differences in specific qualities of particular food products, e.g., the mint flavour of a chewing gum.

Although sensory analysis is often the ultimate test for the acceptance or rejection of a particular food product, there are a number of disadvantages: it is time consuming and expensive to carry out, tests are not objective, it cannot be used on materials that contain poisons or toxins, and it cannot be used to provide information about the safety, composition or nutritional value of a food. For these reasons objective analytical tests, which can be performed in a laboratory using standardised equipment and procedures, are often preferred for testing food product properties that are related to specific sensory attributes. For this reason, many attempts have been made to correlate sensory attributes to quantities that can be measured using objective analytical techniques, with varying degrees of success.

Analytical Technique

There are usually a number of different analytical techniques available to determine a particular property of a food material. It is therefore necessary to select the most appropriate technique for the specific application. The analytical technique selected depends on the property to be measured, the type of food to be analysed, and the reason for carrying out the analysis. Information about the various analytical procedures available can be obtained from a number of different sources. An analytical procedure may already be routinely used in the laboratory or company where you are working. Alternatively, it may be possible to contact an expert who could recommend a certain technique, e.g., a University Professor or a Consultant. Often it is necessary to consult scientific and technical publications. There are a number of different sources where information about the techniques used to analyse foods can be obtained:

Tabulated Official Methods of Analysis

A number of scientific organisations have been setup to establish certain techniques as official methods, e.g. Association of the Official Analytical Chemists (AOAC) and American Oil Chemists Society (AOCS). Normally, a particular laboratory develops a new analytical procedure and proposes it as a new official method to one of the organisations. The method is then tested by a number of independent laboratories using the same analytical procedure and type of equipment stipulated in the original proposal.

The results of these tests are collated and compared with expected values to ensure that the method gives reproducible and accurate results. After rigorous testing the procedure may be accepted, modified or rejected as an official method. Organisations publish volumes that contain the officially recognised test methods for a variety of different food components and foodstuffs. It is possible to consult one of these official publications and ascertain whether a suitable analytical procedure already exists or can be modified for your particular application.

Journals

Analytical methods developed by other scientists are often reported in scientific journals, e.g., Journal of Food Science, Journal of Agriculture and Food Chemistry, Journal of the American Oil Chemists Society, Analytical Chemistry. Information about analytical methods in journals can often be obtained by searching computer databases of scientific publications available at libraries or on the Internet.

Equipment and Reagent Suppliers

Many companies that manufacture equipment and reagents used to analyse foods advertise their products in scientific journals, trade journals, trade directories, and the Internet. These companies will send you literature that describes the principles and specifications of the equipment or test procedures that they are selling, which can be used to determine the advantages and limitations of each technique.

Role of Internet

The Internet is an excellent source of information on the various analytical procedures available for analysing food properties. University lecturers, book suppliers, scientific organisations, scientific journals, computer databases, and equipment and reagent suppliers post information on the web about food analysis techniques. This information can be accessed using appropriately selected keywords in an Internet search engine.

Developing a New Technique

In some cases there may be no suitable techniques available and so it is necessary to develop a new one. This must be done with great care so as to ensure that the technique gives accurate and reliable measurements. Confidence in the accuracy of the technique can be obtained by analysing samples of known properties or by comparing the results of the new technique with those of well-established or official methods.

One of the most important factors that must be considered when developing a new analytical technique is the way in which "the analyte" will be distinguished from "the matrix". Most foods contain a large number of different components, and therefore it is often necessary to distinguish the component being analysed for ("the analyte") from the multitude of other components surrounding it ("the matrix"). Food components can be distinguished from each other according to differences in their molecular characteristics, physical properties and chemical reactions:

- Molecular characteristics: Size, shape, polarity, electrical charge, interactions with radiation.
- Physical properties: Density, rheology, optical properties, electrical properties, phase transitions (melting point, boiling point).
- Chemical reactions: Specific chemical reactions between the component of interest and an added reagent.

When developing an appropriate analytical technique that is specific for a particular component it is necessary to identify the molecular and physicochemical properties of the analyte that are sufficiently different from those of the components in the matrix. In some foods it is possible to directly determine the analyte within the food matrix, but more often it is necessary to carry out a number of preparatory steps to isolate the analyte prior to carrying out the analysis. For example, an analyte may be physically isolated from the matrix using one procedure and then analysed using another procedure.

In some situations there may be one or more components within a food that have very similar properties to the analyte. These "interferents" may make it difficult to develop an analytical technique that is specific for the analyte. It may be necessary to remove these interfering substances prior to carrying out the analysis for the analyte, or to use an analytical procedure that can distinguish between substances with similar properties.

Selecting an Appropriate Technique

Some of the criteria that are important in selecting a technique are listed below:

— *Precision*: A measure of the ability to reproduce an answer between determinations performed by the same scientist (or group of scientists) using the same equipment and experimental approach.

— *Reproducibility*: A measure of the ability to reproduce an answer by scientists using the same experimental approach but in different laboratories using different equipment.

— *Accuracy*: A measure of how close one can actually measure the true value of the parameter being measured, e.g., fat content, or sodium concentration.

— *Simplicity of operation*: A measure of the ease with which relatively unskilled workers may carry out the analysis.

— *Cost*: The total cost of the analysis, including the reagents, instrumentation and salary of personnel required to carry it out.

— *Speed*: The time needed to complete the analysis of a single sample or the number of samples that can be analysed in a given time.

— *Sensitivity*: A measure of the lowest concentration of a component that can be detected by a given procedure.

— *Specificity*: A measure of the ability to detect and quantify specific components within a food material, even in the presence of other similar components, e.g., fructose in the presence of sucrose or glucose.

— *Safety*: Many reagents and procedures used in food analysis are potentially hazardous e.g. strong acids or bases, toxic chemicals or flammable materials.

— *Destructive/Nondestructive*: In some analytical methods the sample is destroyed during the analysis, whereas in others it remains intact.

— *On-line/Off-line*: Some analytical methods can be used to measure the properties of a food during processing, whereas others can only be used after the sample has been taken from the production line.

— *Official Approval*: Various international bodies have given official approval to methods that have been comprehensively studied by independent analysts and shown to be acceptable to the various organisations involved.

— *Nature of Food Matrix*: The composition, structure and physical properties of the matrix material surrounding the analyte often influences the type of method that can be used to carry out an analysis, e.g., whether the matrix is solid or liquid, transparent or opaque, polar or non-polar.

If there are a number of alternative methods available for measuring a certain property of a food, the choice of a particular method will depend on which of the above criteria is most important. For example, accuracy and use of an official method may be the most important criteria in a government laboratory which checks the validity of compositional or nutritional claims on food products, whereas speed and the ability to make nondestructive measurements may be more important for routine quality control in a factory where a large number of samples have to be analysed rapidly.

References

Covello VT, *Merkhofer MW Risk assessment methods: approaches for assessing health and environmental risks*, Plenum Press, New York, 1993.

Kramer, A. and Twigg, B.A., *Fundamentals of quality control for the food industry*, AVI Publishing Co., Westport, Conn. USA, 1962.

2

Quality Control in Food Industry

Quality control is the sum of all those controllable factors that ultimately influence positively or negatively the quality of the finished product e.g. selection of raw materials, processing methods, packaging, methods of storage distribution etc. Quality is defined as any of the features that make something what it is or the degree of excellence or superiority. The word "quality" is used in various ways as applied to food.

Quality product to the salesman means one of high quality and usually at an expensive nature e.g. champagne is considered a quality as compared to fish and chips. Likewise for fresh produce, the word "quality" refers to the attributes of the food which make it agreeable to the person who eats it. This involves positive factors like colour, flavour, texture and nutritive value as well as the negative characteristic such as freedom from harmful microorganisms and undesirable substances.

The term control does not imply that a poor raw material can be converted into a good finished product. In food processing, the general rule is that the effective methods must be carefully applied to conserve the original qualities of the raw materials. Processing cannot improve the raw material. The aim of quality control is to achieve as good and as consistent a standard of quality in the product being produced as is compatible with the market for which the product is designed.

Importance of Quality Assurance

Small and medium sized food processing businesses all over the world increasingly have to consider the production of good quality products as essential to their survival. Consumers and buyers are becoming more aware of the importance of safe, high quality products. Large companies that can afford advertising space on the radio, television or in the press emphasise the quality of their goods, often in a very subtle way. This quality image is given by stating for example "our foods are made only from high quality ingredients."

They also project a quality image through packaging etc. Producers who sell intermediate products, such as dried fruits, to a secondary processor will find that the buyer expects the foods to meet an agreed standard. In the case of exporters, these standards are becoming more and more strict. In order to improve and control product quality it is essential to fully understand the meaning of the term quality. A common definition is "achieving agreed customer expectations or specifications".

The following examples using baked goods illustrate the difference between quality control and quality assurance. A customer may specify that bread should be white, with a good loaf volume and pleasant flavour and taste. The manufacturer then needs to focus on the process to ensure that the raw materials are consistently handled to produce uniform white loaves with the expected volume and taste.

Controlling quality may be achieved by:

— Inspection of raw materials to ensure that no poor quality ingredients are used.

— Carrying out checks on the process to ensure that the weights of the ingredients and temperature and time of baking are correct.

— Inspecting the final product to ensure that no poor quality loaves are sent to the consumer.

However, this Quality Control approach is focused on the process whereas the problems that customers may face can also occur elsewhere in the production and distribution chain. The following examples highlight the shortcomings of a simple quality control approach.

Problem:	Many of the loaves are contaminated with pieces of wood.
Reason:	The distribution system involves transporting the loaves on wooden trays to retail stores where the loaves are packaged and then sold to customers. The wooden trays are not part of the quality control system because they are used after the product has left the bakery.
Problem:	A particular customer has asked for loaves of a different size and colour but these do not arrive as requested.
Reason:	The sales staff have no formal procedure for informing the production staff about changes in this customer's specification. The problem has occurred because of missing links in quality management in the bakery.
Problem:	Bread has been returned because of a bad flavour and some customers have complained that they have been made ill.
Reason:	The flour has been stored next to cleaning chemicals in the dry goods store. One old unlabelled chemical container has been found to have leaked. The company have no documented rules for the storage and handling of chemicals. The staff who routinely clean the store are not trained and receive lower wages than other members of the production team. The container is old and unlabelled.

In order to overcome the types of problems outlined above, a wider approach than quality control is required. This is termed Total Quality Assurance. Quality Assurance systems take a much wider view of what is involved in satisfying customers' needs. The quality assurance system focuses on the prevention of problems and not simply on their cure. Curing problems is expensive and quality cannot be 'inspected into' a product. A quality assurance approach therefore, includes the whole production and distribution system, from the suppliers of important raw materials, through the internal business management to the customer. Quality assurance systems should be documented in a simple way to show who has responsibility for doing what and when. The focus of quality assurance is prevention and this should mean that action is taken to meet a specification and prevent failures from occurring a second time. This is done by planning, management action, agreements with key suppliers and other people in the distribution chain.

Quality assurance can only be operated when staff are well trained and motivated. Workers are normally well aware of the causes of most problems and when quality assurance is used properly they can resolve most quality problems within their control. It is the responsibility of business owners to ensure that the quality assurance system, together with any necessary equipment and information, are available to the workers to allow them to exercise this control.

It is important to recognise that any system is operated by people. It is people who manufacture a food product and ensure that it has the right quality. People working together ensure that the information, materials and equipment are all correct to allow the production of a product. People also store the product and deliver it on time. All therefore need the necessary training and skills to complete their tasks correctly. They need to know what their own responsibilities are in this quality chain and where they fit into the overall system.

Business owners must not regard communication as a one-way process. The information they send to workers must also be modified by feedback from the staff. Well trained and informed staff are an essential element of the Quality Assurance approach.

Analysing Food Processes

All food companies, including the smallest manufacturers, have a responsibility to provide consumers with safe, wholesome foods. Safety is not an option but it is an essential part of the planning, preparation and production of foods.

Any lack of consideration of safety can result in a serious threat to public health. This is recognised by the law in most countries and serious penalties exist for those who contravene hygiene and food safety legislation. At present in many countries, enforcement staff are not always sufficiently resourced to be fully effective, but this situation is changing as consumers become more concerned about food safety.

An important management method to ensure the safety of foods is the Hazard Analysis Critical Control Point (HACCP) system. This is based on quality control, microbiology and risk management and it has been adopted throughout the world, although some countries have tailored the approach to the needs of their particular food sectors. Many small producers may consider that the development of HACCP systems is not feasible or appropriate to their current needs. However larger manufacturers and producer groups who export to industrialised countries are increasingly finding that HACCP is not a matter of choice but is demanded by the importing company. With time, it is likely that the use of HACCP will be more broadly required by food manufacturers.

An alternative approach is for the processor to carefully examine every stage in a process to see where and how improvements can be made in the quality and safety of the finished products. The aim of such an exercise is to focus the attention of operators and the manager on the prevention of problems rather than cures, by identifying potential hazards or quality failures and then developing preventative measures for their control. To implement such an analysis it is necessary to first decide whether the work will focus on improvements to product quality or improvements to the safety of foods. These are obviously connected as food safety is one aspect of quality, but they should be treated as separate exercises.

The level of risks is then assessed and procedures are implemented to monitor and control these risks. If analysis of safety is selected, it is necessary to identify the hazards in a process. This is especially true for high-risk foods (those that can support the growth of food poisoning microorganisms). If high-risk foods are involved then the severity of the hazard is greater and these food products must be investigated thoroughly as very stringent controls are needed. It is for this reason that inexperienced producers should not be encouraged to make high-risk foods such as canned meats and fish.

If quality improvement is selected, it is necessary to identify where a loss in quality is likely to occur in the raw materials or the process and then find methods to control the procedures that are used in order to improve quality in the finished product. Most small-scale processors do not have the necessary skills or time to conduct such a study, and it is likely that assistance will be needed from other people who have the necessary experience. Ideally a small team of people should be assembled to effectively analyse the process and then develop and implement the improvements for the selected product. Those selected should have appropriate expertise of the product, the processing operations, microbiology and quality control. Staff from a manufacturer's association, the local Bureau of Standards, Government Regulatory Food Control Office, a University, Trading Standards Department or suppliers may be suitable resource people to assist staff from the food business itself. For example if the focus of the group is to improve product safety it should conduct its work in the following way:

Gather Information

— Identify sources and routes of contamination by microorganisms, biological, chemical or physical contaminants. It is easier to select one type of hazard which is most important to the product and complete the analysis. Other hazards should then be reviewed later in decreasing order of importance and added to the plan of action.

— Study the effect of the process on levels of contamination and assess the probability of microorganisms surviving the process to grow in the finished product. This should include all parts of the process, from growing foods or buying ingredients to storage and consumption of the finished product.

— The production process for the food is then shown diagrammatically by constructing a process flow diagram, which is taken to the processing building and checked for accuracy.

Find Solutions

— The team then prepares a diagram showing the hazards that have been identified and where they occur in the process. This should be the focal point for discussions by the team.

— The severity of risk of each of the above hazards should be assessed and any areas of doubt should be investigated further.

This should be considered in relation to microorganisms of concern. Stages in the process that destroy, reduce or allow survival of microorganisms should be evaluated.

Implementation

— The owner or manager of the business also needs to be fully convinced that the new procedure is necessary, or else it will fall into disuse after a short time. Operators need to be involved and made aware of problems so that they will understand why new procedures are introduced.

— Control procedures are introduced at the control points using a "decision tree" to help in this procedure. Target limits and tolerances are given for each critical control point. Staff are trained how to operate the new methods and the limits that are placed on any variation from the specified methods.

Monitoring and Documentation

— The group sets up a procedure to monitor the changes that have been introduced and to ensure that everyone involved in the process understands his or her responsibilities. A system for monitoring these is produced, together with a plan for corrective actions, should the tolerances be exceeded.

— Where action is taken it should be clear who has the authority to make decisions and who is responsible for checking that the action was properly done. These responsibilities should be discussed and written down so that everyone is aware of each other's part in the new system. The system is checked and reviewed each year.

When the focus is on quality improvement, a similar sequence of events is used. The quality problems are first identified and possible causes and solutions are discussed. If the problem is associated with poor quality raw materials or ingredients for example, this should be negotiated with suppliers and if needed, quality testing methods can be introduced with agreed tolerance limits. If the problem is associated with the process, for example a critical heating time, then improved process control measures are introduced. In all cases staff must be fully involved and trained so that the improved quality management is sustainable. The effects of any changes should be carefully monitored and recorded.

Such systems need not be complex. Only limited documentation is required and this should assist the small scale processor rather than prevent flexible working. These simple systems are designed to control the key parts of the process and help producers to concentrate valuable manpower where it is most effective.

Design of Buildings Equipment Layout

The type of building in which food products are manufactured and the general level of plant hygiene have a major influence on product quality. Ideally a food manufacturer should have a building constructed specifically for the purpose, but in reality this rarely happens and an existing building has to be modified. If care is taken in the way that the building is adapted, it adds little extra to the total cost but it ensures that the unit is appropriate for food processing.

Location of the Building

The location of the building is very important but is often ignored at the outset. The site should be on cleared ground, away from sources of insects, rodents or smells. It should have a good supply of potable water and if required, electricity. A road access for bringing in raw materials and packaging, and sending out products is usually essential.

Appearance of the Building

The external appearance of the building is a key factor that can influence customers to believe that the company has good management. Externally and internally the building should be clean and painted, with a professionally made nameplate. Ideally the surrounding area should be planted with grass, as short grass acts as a very efficient trap for airborne dust. Washing and toilet facilities must be provided, preferably in a separate building. If this is not possible there must be two closed doors between the toilet and the processing area to prevent insects and odours from entering.

All internal walls of the building should be smooth plastered and painted with a water-resistant paint so that they can be washed. Ideally walls should be tiled to about one to one point five metres above ground level. If this is too expensive then tiling should be carried out around sinks and on walls where food may be splashed. The bottom of the wall, where it meets the floor, is often forgotten. A right angle joint is difficult to clean and can collect dirt. The concrete floor should be curved up to meet the wall and so provide a smooth surface that is easily cleaned. Similarly window ledges should slope so that they do not collect dust, dirt or old cloths that may be left there by workers.

Most manufacturers are aware that windows should be fitted with fly-proof mesh, but they often forget other points through which insects, birds and rodents can enter the processing room or storeroom. Important areas are gaps where the roof meets the walls and gaps in the roof. Rats are also able to get into buildings along power lines and these should be fitted with metal discs at least twenty five centimetres in diameter.

In tropical climates, a large overhanging roof shades the walls, making working conditions better and providing a useful area for activities such as bottle washing. Most types of food processing involve the use of large amounts of water, and floors must be designed so that they drain efficiently. The best way to do this is to slope all floors to a central drainage channel. The drain should be covered with a removable grating to allow cleaning. Drains are a favourite entry point for pests such as rats and cockroaches and the outlets must be fitted with a removable fine mesh.

All electric power points should be fixed at least one to one point five metres high on the walls to keep them dry. Any 3-phase equipment should be installed by a competent electrician. Although they are expensive, waterproof power points are preferred in wet areas. Fluorescent tubes provide good lighting for general work but it must be remembered that normal bulbs should be used near to machines with fast moving parts. This is because fluorescent light can cause a rotating machine to appear stationary at certain speeds; an obvious hazard to workers.

Many food processing operations involve heating, often with the production of steam. Good ventilation is therefore essential and large mesh covered windows, roof vents and ceiling fans should be used. Extraction fans may be required above boiling pans if the heat cannot be removed from the room by other means. Water quality is essential to quality assurance and food processing requires a good supply of clean, potable water for cleaning equipment, cooling filled containers and sometimes as a food ingredient. In many parts of the world the main water supply is unreliable and the manufacturer must use other methods to overcome water supply problems.

Equipment and Layout

Poor equipment layout can be blamed for many quality problems in food factories.

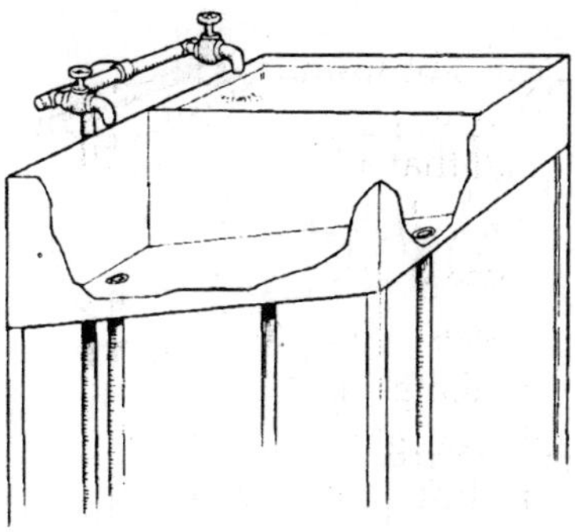

Figure 1. Double tank water purification system

In many small factories workers can be seen almost working against each other, colliding and dropping things. However, good planning and risk assessment can be used to avoid many such errors. The two broad principles to remember are:

— There should be smooth flow of materials around the processing room, from incoming raw materials to finished products.

— Cross-contamination should be avoided. It is easy for example, for spray from unprocessed foods to enter a container of product being filled after processing. This results in contamination and wasted food.

Sanitation and Hygiene

The building and equipment must be kept clean at all times as part of a planned quality assurance programme. A thorough clean-down at the end of the day is essential but this alone is not sufficient. Workers must also be trained to keep equipment clean throughout the day and to remove wastes from the building as they accumulate. The type and frequency of cleaning depends on the food being processed. The most important point is that the manager identifies all areas of potential hazard, then develops a cleaning plan and makes sure all staff are trained and know their particular responsibilities.

Most importantly, the manager should allow adequate time for cleaning down. Too often the final clean-down is carried out in a rush during the last few minutes of the day. In dry processing, or in processes that use dry powders such as flour, it is essential that all dust is cleaned from the building, not forgetting high window ledges, old sacks etc. The objective is to prevent any areas collecting dust where insects can breed.

For wet processes, cleaning involves the use of both detergents and sterilants. Detergents remove food residues but do not kill microorganisms. Sterilants (mainly chlorine) kill microorganisms but do not remove residues. Therefore for good cleaning the residues are first removed with a detergent and the equipment is then treated with a sterilant. A large range of detergents is available for different uses. Manufacturers should investigate which types are available locally and consult suppliers to find the best type for their process.

Chlorinated Water

Chlorinated water is needed for cleaning, washing raw materials and as an ingredient in some products. The required level of chlorine depends on the use of the water. Water for cleaning requires a high level of chlorine; up to 200 ppm. Water used in a product should contain about 0.5 ppm to avoid giving a chlorine flavour to the product. There are a number of ways of preparing chlorinated water for use in the processing plant. The simplest way is to use household bleach, which is readily available in most countries. Table 1 shows the dilutions needed to give different chlorine concentrations.

Table 1. Dilution of bleach

Amount of bleach (ml)	Amount of water (litres)	Chlorine concentration (ppm)
1000	250	200
500	250	100
25	250	5
2.5	250	0.5

Alternatively bleaching powder can be used. This is cheap and when fresh, contains 33 % chlorine. The powder does however weaken with time and tins should always be kept sealed when not in use. Bleaching powder is used by making up a 1% chlorine solution (30-40g per litre). This is then diluted for use: for example 6mls in 45 litres of water gives a 1ppm solution. In cases where a pumped water supply from a well is being used it is common practice to use the action of the pump to automatically dose the water supply with a strong chlorine solution.

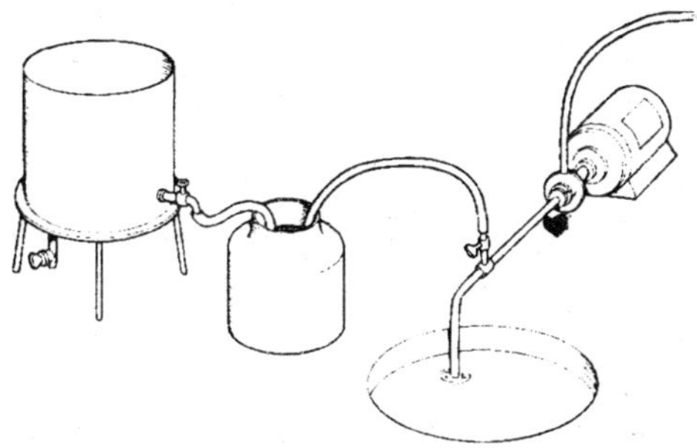

Figure 2. A typical chlorine dosing installation

Although chlorine has the great advantage of killing a wide variety of microorganisms, it also has several disadvantages: it may corrode equipment, particularly aluminium; it can leave flavour taints if it is not rinsed well with potable water; pure bleach Must be Handled with care as it can damage the skin and particularly the eyes, as well as cause breathing difficulties. In addition water in many rural areas of developing countries may be slightly cloudy. This suspended material can be cleared by allowing the water to stand for a few hours before use. A double chambered tank combining settling and chlorination can be fitted onto the roof of a building.

The tank should have a sloping bottom and be fitted with drain valves at each end. Both tanks are filled at the end of the day and chlorine solution is added at the required level.

By the morning the suspended solids will have settled. In use, water is drawn off from the high level valve and when the tank is almost empty the low level valve is opened to flush out any settled material. The tank is then refilled and the water allowed to settle while the second chamber is used.

Operator Hygiene

Operators are a potential source of contamination of foods but to a considerable extent the risk depends on the type of food products. For example, a food that is hot-filled into a bottle, sealed and then heat-treated carries a far smaller risk than a baked meat pie which is handled after it has been cooked. The manager should evaluate the risk and ensure that hygiene procedures are established in the factory and that they are appropriate to the types of products being made. Such measures include the following:

- All workers should use clean uniforms, shoes and hats that cover the hair.
- All workers should scrub their hands and fingernails with unscented soap at the start of each production session. Clean towels or disposable paper towels should be provided.
- Workers should be trained to understand the importance of good hygiene. Local public health departments can usually provide training, posters etc. The training will include:
- Hands should always be washed after using the toilet.
- Smoking and spitting should be banned from the processing rooms.
- No food (including the products) should be eaten in the processing room.
- If affordable, showers should be provided and an area to change clothing.
- Workers who are ill, and especially if suffering from diarrhoea or skin infections should not under any circumstances, be allowed to handle foods.

- Workers with infected cuts, boils or abrasions on their hands should be removed from the production area.

It is very important that workers do not get penalised for having an infection, otherwise they will tend to hide their problem. They should be found other duties; for example, there is always a backlog of cleaning, painting and repairs to be carried out in a production unit. The task of a good manager is to ensure that staff are aware of the risks associated with infections and in this connection wall posters of good and bad practices are very useful. Some of these are available in blank form so that a message can be written in any local language.

Cleaning Schedules

The overall cleanliness of a food processing unit, however small, can have a major impact on the quality of finished products. Cleaning schedules should be seen as an integral part of an overall quality assurance system and a responsible producer carries out a HACCP analysis of cleaning. Areas of hazard need to be identified, the severity of risk evaluated and cleaning procedures put into place. All areas need attention but some carry a greater risk than others.

Each worker should know their cleaning responsibilities within the overall schedule, The owner must take overall responsibility to ensure that cleaning takes place to the correct standard. It is useful to use cleaning cloths and brushes made of brightly coloured materials as these show up easily if they contaminate foods. It is recommended that a cleaning schedule book is maintained which details the area or item to be cleaned, how, when and who is responsible.

Regulatory Quality Control

Almost all countries in the world now have laws governing the production, composition, labelling and safety of processed food and an Agency that is responsible for their implementation. In some countries this agency is the Bureau of Standards in others it may be a division of a Ministry such as Agriculture, Health or Trade and Industry. Although the degree to which national food legislation is currently applied varies greatly from country to country, food manufacturers should be aware that the overall trend is to more stringent application, inspection and control by the responsible authority. National food legislation varies considerably in detail from country to country but applies to three broad areas:

— the plant in which the food is made, its correct design and construction, cleanliness and worker hygiene.

— the physical characteristics of the food (including foreign bodies and adulteration), the chemical composition and microbiological quality.

— the correct labelling of the product including related aspects such as sell-by date, etc.

A food manufacturer contravening national food legislation is subject to penalties which can, in extreme cases, be very severe and involve forced closure of the plant and heavy fines. Manufacturers should be aware that consumers are becoming increasing concerned about food safety and quality issues and are thus more likely to take complaints to the local food control authority. This greatly increases the likelihood of inspection of the premises and products with the resulting risk of penalties.

Food processors should therefore ensure that they know how local legislation applies to their production unit and products. Copies of relevent laws should be obtained and the entire production system should be monitored to ensure that it meets the requirements. Producers that export foods face an even more difficult situation and need obtain copies of relevent food legislation of the importing country as this may vary from their own. In some cases for example, it may be found that a product will need to be reformulated or specially labelled to meet the laws of the importing country.

Unfortunately the authorities charged with the application of food laws are often seen by manufacturers as being threatening and simply "policing" production activities.

Ideally food producers should try to use these authorities as advisers. In many cases they will be able to provide useful guidance to avoid a problem developing.

Simple Methods for Quality Control

The methods described below have been included because:

- They are each relatively simple to use.
- They have sufficient accuracy for quality control.
- They do not require sophisticated or expensive equipment.
- They do not require a high level of skill to operate.
- They are sufficiently inexpensive that they may be used regularly by small food businesses.

Acidity Measurement

To measure the amount of acid (such as citric acid, acetic acid, lactic acid, etc.), it is necessary to titrate a sample of the food with sodium hydroxide solution. It is not sufficient to measure the pH of a food as this does not tell you the amount of acid present.

The method involves the following steps:

- take a 10 ml sample of liquid food or 10 g of solid food
- if solid, the food should be liquidised to a fine pulp
- mix the sample with 90 ml of distilled water, making sure that it is completely mixed
- add about 0.3 ml of an indicator solution
- fill a burette with 0.1 M sodium hydroxide solution (obtainable from larger pharmacies) and titrate the sample until there is a pink colour that does not change.
- calculate the amount of acid as " % acid per ml of liquid food" or "per g of solid food" using the formula: % acid = number of ml of sodium hydroxide x one of the conversion factors below:
    - acetic acid (vinegar) 0.060

— citric acid 0.070
— tartaric acid 0.075
— lactic acid 0.090

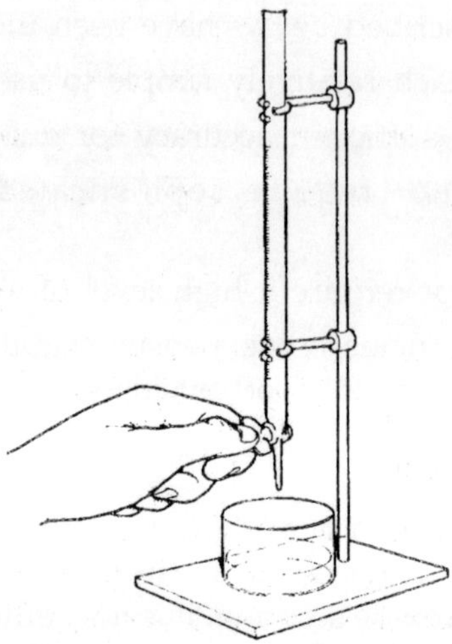

Figure 3. A burette used for titration

It is necessary to know what is the major acid present in a type of food before selecting the conversion factor.

Chlorine Measurement

The determination of levels of chlorine in water is usually carried out using a calorimetric test in which a chemical dye, which reacts with chlorine, produces a colour that is proportional to the amount of chlorine present. The test requires the use of a 'Lovibond Comparator'. This is supplied with a number of discs of coloured glass that are calibrated for different chlorine levels.

In use, a few drops of the dye are placed in test tube. The water under test is added and a colour develops. The tube is then placed in the comparator and the colour is matched with one of the calibrated discs.

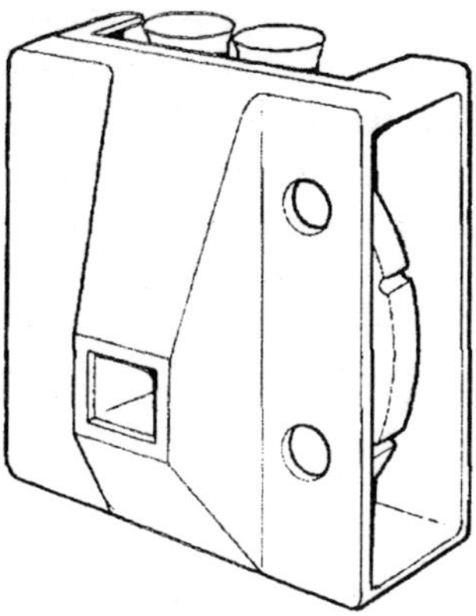

Figure 4. A comparator, used to test chlorine levels in water.

These chlorine test kits are available from a number of suppliers and it is suggested that those considering testing chlorine levels in water should consult the local Water Department which will be able to advise on the nearest source of supply and possibly provide training in the method.

Fill-weight Measurement

In most countries there are legal requirements which state that a container has the weight of its contents written on the label and that the net weight of food inside is not less that this weight. This should be routinely checked by taking regular samples of filled containers and placing them on a scale. On the other side of the scale there is the heaviest empty container from a batch plus a metal weight that is equal to the net weight shown on the label. Filled containers that are underweight should be removed and re-filled. The results of these checks should be recorded on a chart and related back to the worker who is filling the containers or operating a filling machine to ensure that accuracy of filling increases with experience gained.

It should be noted that this is the "minimum weight" method of checking fill-weights. It is intended to ensure that all packs contain more than the net weight shown on the label and it is a simple system for small scale producers to operate.

There is however another system, known as the "average weight" system, which relies on the statistical probability that a known proportion of packs will be above the net weight. This system was developed in Europe for automatic filling operations and is unnecessarily complex for small scale producers. However if exports to Europe, USA or some other countries are contemplated, then it is recommended that details of this system should be obtained from a local Export Development Board or its equivalent.

Flour Infestation

— 100 grams of flour is weighed onto a flat surface.

— Flour is flattened using a ruler.

— Flour is examined after two minutes for evidence of pimpling.

Pimpling indicates the presence of flour mites breaking the surface for air.

Glass Container Measurement

Glass containers have more variable dimensions than either plastic or metal containers because of the nature of their manufacture. It is therefore important to check particular dimensions to ensure that,

— a container has the expected capacity;

— that the neck is properly formed and will allow the lid to fit properly;

— that the container is vertical to prevent it breaking in a filling machine, and

— that the weight of a sample of empty containers is checked to find the heaviest in a batch for use in check-weighing.

Glass pieces are also a particular danger to customers if they become mixed into the product. It is therefore essential that the checks described below are routinely performed on all glass containers. Finally glass containers are often re-used and may have become contaminated by kerosene, pesticides or other materials. They should be thoroughly washed and inspected by looking and smelling to ensure that no residues remain before the food is filled.

Weight of Containers

Take a random sample of empty containers from an incoming batch (for example 1 in 50) and weigh them, together with their lids. The required check-weight is calculated as the weight of the heaviest container plus the net weight of product.

Weigh a dried container and then fill it to the top with distilled water at 20°C. Reweigh the filled container and the difference in weight (in grams) is equivalent to the capacity in ml. This capacity should be great enough to allow sufficient food to be filled to meet the net weight declared on the label.

Headspace Gauge

A space is required between the surface of a hot-filled product in a jar and the underneath surface of the lid. This allows a partial vacuum to form when the product cools and thus helps to prevent spoilage. The volume of the headspace does not normally exceed 10% of the capacity of the container. Measuring the depth of the headspace is a quick method of assessing the volume, but the depth varies according to the capacity of the container (larger containers require a deeper headspace). A headspace gauge is a cheap and simple way of routinely checking that product has been filled to the correct level. It consists of a series of prongs of different lengths fixed onto a bar and it is placed on the rim of the jar before fitting the lid. The level of product may then be seen where it touches one of the prongs.

Partial Vacuum

The partial vacuum in hot-filled glass jars may be measured using a Bourdon tube vacuum gauge. The gauge is fitted with a sharp needle, surrounded by a rubber seal. The needle is pushed through the lid of the container and the moistened rubber seal prevents air from entering. The partial vacuum may then be read directly from the gauge as 'mm of mercury' or 'minus kPa'. As the product is not saleable after this test, it is usually only applied to a small sample or when a problem arises.

Dimensions of Containers

The important routine checks are to measure the height of containers, their neck diameter and outside diameter, and their ovality (to ensure that they are round and not oval). Simple equipment may be manufactured to perform these checks. A vertical ruler on a stand is used to measure height. Different "go/no-go" rings can be made for each size of container that is used. Rings are slipped over the neck of a container to quickly show whether the diameter is too large or too small for the intended lid and also to show if the neck is not circular. Different sized rings can similarly be used to check the outside diameter of the container and its ovality.

Faults in Glass

It is essential that all glass containers are visually checked to make sure that no glass splinters or cracks are present. Common faults in glass are bubbles, cracks and strings. A light-box, in which a light bulb is placed behind a translucent plastic screen, is useful to view glass containers clearly. Operators who check glass containers should be fully trained in which faults to look for and they should be moved from inspection after 3060 minutes to prevent tiredness and lack of concentration.

Gluten Measurement

- Weigh 10 + 0.01 g of flour and place in a basin.
- Add 6ml of water to the basin, (5 ml will be sufficient for weak flour).
- Using a spatula, mix the flour and water into a dough. Form the dough into a round ball by rolling between the palms of the hands.
- Replace the dough in the basin and cover it with water.

Leave for a time, at least 10 minutes, preferably 45-60 minutes.

— Holding the dough ball in one hand under cold running water, wash out the starch. Squeeze the dough frequently between the fingers and the palm to help the process.

— When all the starch has been removed the wash water will run clear and the remaining gluten will be free from lumps.

— Remove the excess water with blotting paper.

— Weigh the wet gluten and record this as a percentage of the flour weight.

— The gluten may be dried in an oven at 103°C to determine dry gluten.

Label Measurement

It is often forgotten that labels are an integral and important part of a food product. They need to be tested and checked in the same way as any ingredient. On delivery, samples should be taken from each pack of labels for examination as faults may develop in a print run. All packets of labels should then be repacked and sealed.

Label faults may be divided into major and minor faults. Major faults include:

— The use of incorrect colours.

— Major variations in dimensions.

— Information missing, wrongly printed or mix-spelt.

— Major print errors, slippage.

— No glue, if applicable.

Labels showing major faults should not be used and should be returned to the printer.

Minor faults include:

— Detectable but acceptable colour variations.

— Size error but label usable.

— Minor colour registration errors.

Labels with minor faults may be used but the problems should be discussed with the printer and possibly a reduction in price negotiated.

Loaf Volume Measurement

It is usual to use a simple device in which the displacement of rapeseed or mustard seed is measured. This is accurate because the individual seeds are hard and quite round, flow easily, do not disintegrate and a given weight always occupies the same volume. There are two rectangular compartments, connected by a graduated cylinder made of glass or transparent plastic.

The equipment may easily be inverted, allowing either compartment to be uppermost. An adequate amount of seed is placed inside and flows from one compartment to the other as the apparatus is inverted.

With the seed in one compartment in the lower position, the loaf under test is placed in the top compartment which is then closed and the apparatus is inverted so that the seed fills the space around the loaf and levels off in the tube. The bigger the loaf, the higher up the tube will be the surface of the seed layer.

The actual volume is read off from the graduations on the tube, which is previously calibrated. Loaf volume is usually expressed in cc's and the volume of 1 lb loaves may vary between 1,400 cc's and 1,600 cc's, depending on the flour used. An indication of volume may quickly be obtained by measuring the maximum height of the loaf. This is useful but because of the irregular shape involved, it is not very accurate.

Moisture Content Measurement

With experience, an operator may assess the correct moisture content of grains by placing them on a hard surface and tapping them with a metal or stone weight. The hardness (or softness) of the grain indicates the approximate moisture content. A more accurate but more time consuming method is to dry a weighed sample of grain in an oven at 100°C for 5 hours (or 104°C for 2 hours) and re-weigh. Certain items of equipment are needed to determine the moisture content: a balance accurate to three decimal places (ie 0.001 g), a thermostatically controlled oven and a laboratory desiccator. A sample of material is dried to constant weight and the loss reported as moisture content.

Approximately 2 g of the material under test is accurately weighed (to 0.001 g) into a small dish. This is then placed in the oven for 1 hour, removed from the oven and put in the desiccator to cool. It is then weighed. The dish is replaced in the oven for 30 minutes and the process repeated to constant weight.

The moisture content is found using the following formula:

$$\% \text{ moisture} = \frac{(\text{inital weight} - \text{final weight})}{\text{initial weight}} \times 100$$

A faster but more expensive method is to use a moisture meter. This measures the conductance of electricity through a sample of grain to indicate the amount of water it contains. The instrument is expensive and therefore likely to be affordable only by larger scale millers.

Solids Content Measurement

The method involved is the same as that described for moisture content above, but the result is expressed as ' % solids'. This is calculated using the following formula:

$$\% \text{ solids} = \frac{\text{final weight of sample}}{\text{initial weight of sample}} \times 100$$

Packaging Film Measurement

Made-up plastic bags and rolls of film need to be checked and there are a number of simple tests that a small food processor may carry out. It should be remembered that there is no way of checking for faults inside a roll of film as only the outer part may be seen. Rolls thus need to be examined during use. Typical faults in plastic bags and films include:

— *Incorrect yield:* The barrier properties of a given type of film depend on its thickness. The normal way to measure thickness is by weight per square metre. Using a template 10 squares of film, each 10cm by 10cm are cut out. These are then carefully weighed. The result (in grams/square metre) is then checked against the supplier's specifications.

— *Incorrect printing*: This is described in more detail under quality control of labels.

— *Odour:* Some films are manufactured using solvents and rolls should be checked for any such smell by crumpling a sample in the hand and smelling it.

— *Blocking*: This fault results from layers of film on a roll sticking together. Blocking rarely causes serious problems unless automatic packaging machines are used.

— *Seal strength:* Samples of film should be heat sealed and the seal strength checked by pulling the seal apart. The same test should be applied to made-up bags.

— *Curl:* This causes the film to curl up rather than lay flat. It is caused by the film being poorly stored, particularly in conditions that are too damp or too dry.

pH Measurement

pH is a scale that is used to describe acidity (pH 1-6), neutrality (pH 7) or alkalinity (pH 814). There are two methods of measuring the pH of a sample of liquid food: the simplest and cheapest is to dip a piece of pH paper into the sample.

The paper is impregnated with chemicals that change colour and the colour may be compared to a chart supplied with the paper to give the pH of the sample. This method is often sufficiently accurate for routine Q checks.

If greater accuracy is required a pH meter should be used. These may be mains powered bench models or battery powered portable models. In general bench models are more accurate than portable types, although newer equipment has reduced this difference.

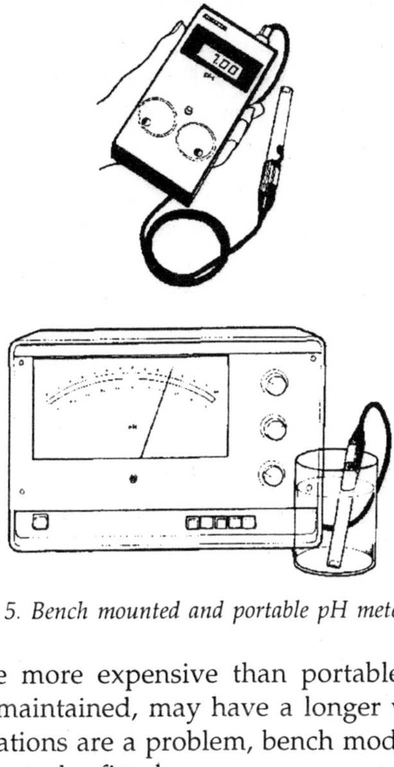

Figure 5. Bench mounted and portable pH meters

Bench types are more expensive than portable types and, when properly maintained, may have a longer working life. If voltage fluctuations are a problem, bench models require a voltage regulator to be fitted.

Modern portable pH meters are fitted with a container filled with buffer solution when they are delivered. This should be replaced as directed in the suppliers instructions. New electrodes for bench models should be soaked for several hours in distilled water or buffer solution.

Afterwards they should be stored with their tips in one of these solutions. Older electrodes may be cleaned by placing in 0.1 M sodium hydroxide solution for 1 minute and then in 0.1 M hydrochloric acid for 1 minute, repeated twice. They are then rinsed in water and carefully blotted (not wiped) with a soft cloth or tissue paper.

pH meters should be standardised against buffer solutions which have a known pH. The standardisation and later pH measurements should all be done at the same temperature (between 20-30°C) to avoid errors in the results.

In use the instructions supplied with the equipment should be carefully followed. For example the instrument has a temperature compensation control that should be set to the local ambient temperature.

The general procedure for measuring pH is as follows:

1. Standardise the instrument using a commercially supplied pH 4.0 buffer or freshly made 0.05 M potassium acid phthalate solution (10.12 g of the chemical in 1 litre of distilled water). Dip the electrode into the buffer and adjust the standardisation control so that the scale reads pH 4.0.
2. Rinse the electrode in water and carefully blot (not wipe) clean with a soft tissue.
3. Check the pH of a commercially supplied pH 7.0 buffer or 0.025 M potassium dihydrogen phosphate solution (3.387 g of the chemical dissolved in 1 litre of distilled water).
4. Repeat step 2.
5. Place the electrode in a sample at the same temperature as the buffers (between 20-30°C) and allow the instrument to stabilise for 1 minute. Repeat step 2 and take a second reading. The two results should be within 0.1 pH units of each other.
6. Store the cleaned electrode in distilled water or buffer solution and switch the instrument to 'stand by' when it is not being used.

Plastic Container Measurement

There are fewer checks that are needed on plastic containers, compared to glass containers. This is because the method of manufacture results in more uniform dimensions, the weight of the container is small compared to glass and variations are therefore less important.

The main faults are likely to be splits, punctures, a badly formed neck and the use of non food-grade plastic. With the exception of the last fault, each can be checked visually by operators involved in filling the containers.

There are no simple checks to ensure that a container is made from food grade plastic and if there is any doubt the processor should consult a reputable supplier for advice. They may be checked to ensure that the seal or cap is watertight by simply inverting a sample of filled containers to detect leaks.

Salt Measurement

There are three methods that can be used for measuring the salt concentration in foods: hydrometry, refractometry or salt titration. Refractometers are expensive and the titration method is more complex and requires training and laboratory chemicals. Hydrometers are hollow glass rods with a bulb at one end. They are sealed at both ends so that they float when immersed in a liquid.

The bulb is weighted so that the hydrometer partially sinks to a level that depends on the specific gravity of the brine (the more salt that there is in a solution, the higher the specific gravity). A scale on the stem of the hydrometer is calibrated and may read from 0-100 degrees, where 0 is pure water and 100 is saturated salt solution (26.5%).

It is important that the measurements are made at the reference temperature for the hydrometer (usually 20°C) because the specific gravity of the brine changes at different temperatures.

The method of measurement involves placing a sample of brine at the correct temperature into a large clear glass or plastic cylinder and gently lowering the hydrometer into the liquid. When it has stopped moving, the scale is read at the surface of the liquid and the reading is converted to % salt using a conversion table supplied with the hydrometer. It is important that a salt hydrometer is specified as there are other types that are calibrated for alcohol or for sugar solutions.

Sieving Tests

A 500 g sample of flour or ground spice is sieved through a stack of metal sieves with the largest mesh at the top of the stack and the smallest at the base. Typically the range of sieve aperture sizes is 1.6 mm to 0.038 mm.

The sieves may be placed on a shaker to achieve a consistent amount of shaking. The amount of material that is collected on each sieve is weighed and expressed as a percentage of the total weight. This method can also be used to detect gross contamination with stalks, stones, string, cigarette ends, leaves etc as these are retained on the larger aperture sieves and can be examined, recorded or weighed.

Filth Test

This is a modified method that allows detection of insect parts, rodent hairs or ground faeces in milled spices or flours. The sample of food is mixed with petrol and thoroughly stirred. The insect parts, etc. are preferentially wetted by the petrol and when the suspension of particles settles, these may be seen floating on the surface of the petrol. If required they may be filtered through a filter paper and examined or identified.

Sodium Benzoate Measurement

Although it is possible to measure the amount of sodium benzoate in a food by measuring the benzoic acid content, this is a fairly complex method that requires laboratory facilities and it is unlikely to be routinely done by a small scale producer.

Sodium Metabisulphite Measurement

The amount of sulphur dioxide that is produced from sodium metabisulphite is approximately two thirds. For example if 1.5 g of sodium metabisulphite is added to one litre of juice it will form 1 g of sulphur dioxide. (la per litre = 0.001%. This is equivalent to 1000 ppm). The amount of sulphur that is required to produce sulphur dioxide in a sulphur cabinet is usually only estimated approximately because of the large number of variables that influence the absorption of sulphur dioxide by fruits.

As an approximate estimate, 350400 g of sulphur can be used per 100 kg of fresh fruit. Although it is possible to measure the sulphur dioxide content of a food item, this requires relatively sophisticated laboratory equipment and is not usually done by small scale producers.

Starch Gelatinisation Measurement

- 100 g of flour is mixed with 900 g of hot water in a pot.
- The pot is heated until the flour mixture has gelatinised.
- The mixture is poured into a 1 litre measuring vessel.
- The measuring vessel is stood in hot water.
- A steel ball is dropped into the mixture and the time to drop 200 ml recorded.
- The time (in seconds) is compared against the standard batch.

Sugar Measurement

Fruit jams, juices, sauces, confectionery, etc. contain sugar as the main soluble solid. For these products the sugar content can be measured directly using a refractometer.

Table 2. Boiling temperatures of different sugar syrup concentrations

Sugar concentration (% sucrose)	Boiling point (°C)
40	101.4
50	102
60	103
711	105.5

Although this equipment is relatively expensive for a small scale producer, it does give an accurate measurement of sugar concentration which is a vital control point for many products. Two types of refractometers are available: the bench type and a hand held type. For quality control purposes the hand held type is cheaper and it is usually sufficiently accurate.

The method involves taking a small sample of the food and placing it on the lower glass prism of the instrument. The upper prism is then closed and the refractometer is held against the eye, pointing in the direction of a window or bright light. It is focused until the scale can be read against a clearly defined division between black and orange colours. The reading is recorded as degrees Brix which corresponds to % sucrose.

Simple sugar syrups may also be measured using hydrometry. The hydrometers are similar to those described for salt, but they are calibrated for sugar (sucrose). The method used is the same as that described for salt and the scale is read as % sucrose. The samples should be at the reference temperature for the hydrometer.

The sugar content of products such as jams and confectionery can be estimated in a less accurate way by measuring the temperature of boiling.

As the sugar content increases the temperature of boiling also increases. Note that the boiling temperature also changes according to the amount of invert sugar or glucose syrup in the boiling mixture and experience of making the product is needed before using boiling temperature as a control measure.

The boiling point also changes according to height above sea level and this should be checked if a producer is operating in a mountainous region.

A special thermometer that reads up to 150°C is required and the bulb of the thermometer should be protected by a metal casing to protect it against breaking. In general mercury thermometers should not be used in food premises.

A crude estimate of the solids content of jam and confectionery products may be made by placing a sample on a jar lid which is floating in cold water and noting the texture of the product after it has cooled to see if a firm gel is formed.

Table 3. Simple tests for sugar boiling

Approx Temperature (°C)	Test	Name	Result
103	A	Thread	Thin strands
105	B	Small pearl	Forms small droplets
105	C	Jam set	Forms a strong gel
106	B	Large pearl	Forms large droplets
111	B	Feather	Forms hard feathery strands
116	B	Small ball	Forms soft ball
120	B	Large ball	Forms hard ball
129	B	Light crack	Forms thin sheet
133	B	Medium crack	Sheet forms, slightly brittle
143	B	Hard crack	Sheet forms rapidly
180	B	Caramel	Brown brittle sheet forms

With experience this may be used as a simple check to ensure that products have been boiled to the correct consistency, but a more accurate measurement using a refractometer is recommended to ensure uniform product quality. A summary of the methods and expected results is given in Table 3.

References

Anon, *Procedures to implement HACCP systems*, IAMFES, Published by International Association of Milk, Food and Environmental Sanitarians, USA, 1991.

__________, *Quality Control for the Food Industry*, ITC UNCTAD/GATT, Geneva, Switzerland, 1991.

Bryan, F.1., *Hazard Analysis Critical Control Point Evaluations*, WHO, Geneva, Switzerland, 1992.

__________, *HACCP in street vending in developing countries*, Food Australia, 45. (2), 1993.

Caplen, R.H., *A practical approach to quality control*, Hutchinson Publishing Group, London, UK, 1982.

Hobbs, B.C. and Roberts, D., *Food Poisoning and Food Hygiene*, Edward Arnold, London, UK, 1987.

Kramer, A. and Twigg, B.A., *Fundamentals of quality control for the food industry*, AVI Publishing Co., Westport, Conn. USA, 1962.

Manuals of Food Quality Control, Vols 1-9, FAO Publications, Via delle Terme di Caracalla, 00100 Rome, Italy.

Shapton, D.A., and Shapton, N.F., *Safe Processing of Foods*, Butterworth-Heinemann, Oxford, UK, 1993.

3

Food Sampling and Data Analysis

A food analyst often has to determine the characteristics of a large quantity of food material, such as the contents of a truck arriving at a factory, a days worth of production, or the products stored in a warehouse. Ideally, the analyst would like to analyse every part of the material to obtain an accurate measure of the property of interest, but in most cases this is practically impossible. Many analytical techniques destroy the food and so there would be nothing left to sell if it were all analysed.

Another problem is that many analytical techniques are time consuming, expensive or labour intensive and so it is not economically feasible to analyse large amounts of material. It is therefore normal practice to select a fraction of the whole material for analysis, and to assume that its properties are representative of the whole material. Selection of an appropriate fraction of the whole material is one of the most important stages of food analysis procedures, and can lead to large errors when not carried out correctly.

Population, Samples and Laboratory Samples

It is convenient to define some terms used to describe the characteristics of a material whose properties are going to be analysed.

— *Population:* The whole of the material whose properties we are trying to obtain an estimate of is usually referred to as the "population".

— *Sample:* Only a fraction of the population is usually selected for analysis, which is referred to as the "sample". The sample may be comprised of one or more sub-samples selected from different regions within the population.

— *Laboratory Sample:* The sample may be too large to conveniently analyse using a laboratory procedure and so only a fraction of it is actually used in the final laboratory analysis. This fraction is usually referred to as the "laboratory sample".

The primary objective of sample selection is to ensure that the properties of the laboratory sample are representative of the properties of the population, otherwise erroneous results will be obtained. Selection of a limited number of samples for analysis is of great benefit because it allows a reduction in time, expense and personnel required to carry out the analytical procedure, while still providing useful information about the properties of the population. Nevertheless, one must always be aware that analysis of a limited number of samples can only give an estimate of the true value of the whole population.

Sampling Plans

To ensure that the estimated value obtained from the laboratory sample is a good representation of the true value of the population it is necessary to develop a "sampling plan". A sampling plan should be a clearly written document that contains precise details that an analyst uses to decide the sample size, the locations from which the sample should be selected, the method used to collect the sample, and the method used to preserve them prior to analysis.

It should also stipulate the required documentation of procedures carried out during the sampling process. The choice of a particular sampling plan depends on the purpose of the analysis, the property to be measured, the nature of the total population and of the individual samples, and the type of analytical technique used to characterise the samples. For certain products and types of population sampling plans have already been developed and documented by various organisations which authorise official methods, e.g., the Association of Official Analytical Chemists (AOAC).

Purpose of Analysis

The first thing to decide when choosing a suitable sampling plan is the purpose of the analysis. Samples are analysed for a number of different reasons in the food industry and this affects the type of sampling plan used:

— *Official samples*: Samples may be selected for official or legal requirements by government laboratories. These samples are analysed to ensure that manufacturers are supplying safe foods that meet legal and labelling requirements. An officially sanctioned sampling plan and analytical protocol is often required for this type of analysis.

— *Raw materials:* Raw materials are often analysed before acceptance by a factory, or before use in a particular manufacturing process, to ensure that they are of an appropriate quality.

— *Process control samples*: A food is often analysed during processing to ensure that the process is operating in an efficient manner. Thus if a problem develops during processing it can be quickly detected and the process adjusted so that the properties of the sample are not adversely effected. Techniques used to monitor process control must be capable of producing precise results in a short time. Manufacturers can either use analytical techniques that measure the properties of foods on-line, or they can select and remove samples and test them in a quality assurance laboratory.

— *Finished products:* Samples of the final product are usually selected and tested to ensure that the food is safe, meets legal and labelling requirements, and is of a high and consistent quality. Officially sanctioned methods are often used for determining nutritional labelling.

— *Research and Development:* Samples are analysed by food scientists involved in fundamental research or in product development. In many situations it is not necessary to use a sampling plan in R&D because only small amounts of materials with well-defined properties are analysed.

Type of Property Measured

Once the reason for carrying out the analysis has been established it is necessary to clearly specify the particular property that is going to be measured, e.g., colour, weight, presence of extraneous matter, fat content or microbial count. The properties of foods can usually be classified as either attributes or variables. An attribute is something that a product either does or does not have, e.g., it does or does not contain a piece of glass, or it is or is not spoilt. On the other hand, a variable is some property that can be measured on a continuous scale, such as the weight, fat content or moisture content of a material. Variable sampling usually requires less samples than attribute sampling.

The type of property measured also determines the seriousness of the outcome if the properties of the laboratory sample do not represent those of the population. For example, if the property measured is the presence of a harmful substance (such as bacteria, glass or toxic chemicals), then the seriousness of the outcome if a mistake is made in the sampling is much greater than if the property measured is a quality parameter (such as colour or texture). Consequently, the sampling plan has to be much more rigorous for detection of potentially harmful substances than for quantification of quality parameters.

Nature of Population

It is extremely important to clearly define the nature of the population from which samples are to be selected when deciding which type of sampling plan to use. Some of the important points to consider are listed below:

— *A population may be either finite or infinite*: A finite population is one that has a definite size, e.g., a truckload of apples, a tanker full of milk, or a vat full of oil. An infinite population is one that has no definite size, e.g., a conveyor belt that operates continuously, from which foods are selected periodically. Analysis of a finite population usually provides information about the properties of the population, whereas analysis of an infinite population usually provides information about the properties of the process. To facilitate the development of a sampling plan it is usually convenient to divide an "infinite" population into a number of finite population, e.g., all the products produced by one shift of workers, or all the samples produced in one day.

— *A population may be either continuous or compartmentalised:* A continuous population is one in which there is no physical separation between the different parts of the sample, e.g., liquid milk or oil stored in a tanker. A compartmentalised population is one that is split into a number of separate sub-units, e.g., boxes of potato chips in a truck, or bottles of tomato ketchup moving along a conveyor belt. The number and size of the individual sub-units determines the choice of a particular sampling plan.

— *A population may be either homogenous or heterogeneous*: A homogeneous population is one in which the properties of the individual samples are the same at every location within the material, whereas a heterogeneous population is one in which the properties of the individual samples vary with location. If the properties of a population were homogeneous then there would be no problem in selecting a sampling plan because every individual sample would be representative of the whole population. In practice, most population are heterogeneous and so we must carefully select a number of individual samples from different locations within the population to obtain an indication of the properties of the total population.

Nature of Procedure

The nature of the procedure used to analyse the food may also determine the choice of a particular sampling plan, e.g., the speed, precision, accuracy and cost per analysis, or whether the technique is destructive or non-destructive. Obviously, it is more convenient to analyse the properties of many samples if the analytical technique used is capable of rapid, low cost, nondestructive and accurate measurements.

Develop a Sampling Plan

After considering the above factors one should be able to select or develop a sampling plan which is most suitable for a particular application. Different sampling plans have been designed to take into account differences in the types of samples and population encountered, the information required and the analytical techniques used. Some of the features that are commonly specified in official sampling plans are listed below.

Sample Size

The size of the sample selected for analysis largely depends on the expected variations in properties within a population, the seriousness of the outcome if a bad sample is not detected, the cost of analysis, and the type of analytical technique used. Given this information it is often possible to use statistical techniques to design a sampling plan that specifies the minimum number of sub-samples that need to be analysed to obtain an accurate representation of the population.

Often the size of the sample is impractically large, and so a process known as sequential sampling is used. Here sub-samples selected from the population are examined sequentially until the results are sufficiently definite from a statistical viewpoint.

Sample Location

In homogeneous population it does not matter where the sample is taken from because all the sub-samples have the same properties. In heterogeneous population the location from which the sub-samples are selected is extremely important.

In random sampling the sub-samples are chosen randomly from any location within the material being tested. Random sampling is often preferred because it avoids human bias in selecting samples and because it facilitates the application of statistics. In systematic sampling the samples are drawn systematically with location or time, e.g., every 10th box in a truck may be analysed, or a sample may be chosen from a conveyor belt every 1 minute.

This type of sampling is often easy to implement, but it is important to be sure that there is not a correlation between the sampling rate and the sub-sample properties. In judgment sampling the sub-samples are drawn from the whole population using the judgment and experience of the analyst. This could be the easiest sub-sample to get to, such as the boxes of product nearest the door of a truck.

Alternatively, the person who selects the sub-samples may have some experience about where the worst sub-samples are usually found, e.g., near the doors of a warehouse where the temperature control is not so good. It is not usually possible to apply proper statistical analysis to this type of sampling, since the sub-samples selected are not usually a good representation of the population.

Sample Collection

Sample selection may either be carried out manually by a human being or by specialised mechanical sampling devices. Manual sampling may involve simply picking a sample from a conveyor belt or a truck, or using special cups or containers to collect samples from a tank or sack. The manner in which samples are selected is usually specified in sampling plans.

Laboratory Samples Preparation

Once we have selected a sample that represents the properties of the whole population, we must prepare it for analysis in the laboratory. The preparation of a sample for analysis must be done very carefully in order to make accurate and precise measurements.

Making Samples Homogeneous

The food material within the sample selected from the population is usually heterogeneous, i.e., its properties vary from one location to another. Sample heterogeneity may either be caused by variations in the properties of different units within the sample (inter-unit variation) and/or it may be caused by variations within the individual units in the sample. The units in the sample could be apples, potatoes, bottles of ketchup, containers of milk etc.

An example of inter-unit variation would be a box of oranges, some of good quality and some of bad quality. An example of intra-unit variation would be an individual orange, whose skin has different properties than its flesh. For this reason it is usually necessary to make samples homogeneous before they are analysed, otherwise it would be difficult to select a representative laboratory sample from the sample.

A number of mechanical devices have been developed for homogenising foods, and the type used depends on the properties of the food being analysed. Homogenisation can be achieved using mechanical devices, enzymatic methods or chemical methods.

Reducing Sample Size

Once the sample has been made homogeneous, a small more manageable portion is selected for analysis. This is usually referred to as a laboratory sample, and ideally it will have properties which are representative of the population from which it was originally selected. Sampling plans often define the method for reducing the size of a sample in order to obtain reliable and repeatable results.

Preventing Changes in Sample

Once we have selected our sample we have to ensure that it does not undergo any significant changes in its properties from the moment of sampling to the time when the actual analysis is carried out, e.g., enzymatic, chemical, microbial or physical changes. There are a number of ways these changes can be prevented.

- *Enzymatic Inactivation:* Many foods contain active enzymes they can cause changes in the properties of the food prior to analysis, e.g., proteases, cellulases, lipases, etc. If the action of one of these enzymes alters the characteristics of the compound being analysed then it will lead to erroneous data and it should therefore be inactivated or eliminated. Freezing, drying, heat treatment and chemical preservatives (or a combination) are often used to control enzyme activity, with the method used depending on the type of food being analysed and the purpose of the analysis.
- *Lipid Protection:* Unsaturated lipids may be altered by various oxidation reactions. Exposure to light, elevated temperatures, oxygen or pro-oxidants can increase the rate at which these reactions proceed. Consequently, it is usually necessary to store samples that have high unsaturated lipid contents under nitrogen or some other inert gas, in dark rooms or covered bottles and in refrigerated temperatures. Providing that they do not interfere with the analysis antioxidants may be added to retard oxidation.
- *Microbial Growth and Contamination:* Microorganisms are present naturally in many foods and if they are not controlled they can alter the composition of the sample to be analysed. Freezing, drying, heat treatment and chemical preservatives (or a combination) are often used to control the growth of microbes in foods.

— *Physical Changes:* A number of physical changes may occur in a sample, e.g., water may be lost due to evaporation or gained due to condensation; fat or ice may melt or crystallise; structural properties may be disturbed. Physical changes can be minimised by controlling the temperature of the sample, and the forces that it experiences.

Sample Identification

Laboratory samples should always be labelled carefully so that if any problem develops its origin can easily be identified. The information used to identify a sample includes:

— Sample description,

— Time sample was taken,

— Location sample was taken from,

— Person who took the sample, and,

— Method used to select the sample.

The analyst should always keep a detailed notebook clearly documenting the sample selection and preparation procedures performed and recording the results of any analytical procedures carried out on each sample. Each sample should be marked with a code on its label that can be correlated to the notebook. Thus if any problem arises, it can easily be identified.

Data Analysis and Reporting

Food analysis usually involves making a number of repeated measurements on the same sample to provide confidence that the analysis was carried out correctly and to obtain a best estimate of the value being measured and a statistical indication of the reliability of the value. A variety of statistical techniques are available that enable us to obtain this information about the laboratory sample from multiple measurements.

Measure of Central Tendency of Data

The most commonly used parameter for representing the overall properties of a number of measurements is the mean:

$$\overline{x} = \frac{x_1 + x_2 + x_3 + \ldots x_n}{n} = \frac{\sum_{i-1}^{n} x_i}{n} \qquad (1)$$

Here n is the total number of measurements, x_i is the individually measured values and $\overline{x}$ is the mean value.

The mean is the best experimental estimate of the value that can be obtained from the measurements. It does not necessarily have to correspond to the true value of the parameter one is trying to measure.

There may be some form of systematic error in our analytical method that means that the measured value is not the same as the true value. Accuracy refers to how closely the measured value agrees with the true value.

The problem with determining the accuracy is that the true value of the parameter being measured is often not known. Nevertheless, it is sometimes possible to purchase or prepare standards that have known properties and analyse these standards using the same analytical technique as used for the unknown food samples.

The absolute error Eabs, which is the difference between the true value (xtrue) and the measured value (x_i), can then be determined: Eabs = (x_i - xtrue). For these reasons, analytical instruments should be carefully maintained and frequently calibrated to ensure that they are operating correctly.

Spread of Data

The spread of the data is a measurement of how closely together repeated measurements are to each other. The standard deviation is the most commonly used measure of the spread of experimental measurements. This is determined by assuming that the experimental measurements vary randomly about the mean, so that they can be represented by a normal distribution. The standard deviation SD of a set of experimental measurements is given by the following equation:

$$SD = \sqrt{\frac{\sum_{i=1}^{n}(x_i - \bar{x})^2}{n-1}} \qquad (2)$$

Measured values within the specified range:

$\bar{x} \pm SD$ *means 68% values within range (x - SD) to (x + SD)*

$\bar{x} \pm 2SD$ *means 95% values within range (x - 2SD) to (x + 2SD)*

$\bar{x} \pm 3SD$ *means >99% values within range (x - 3SD) to (x + 3SD)*

Another parameter that is commonly used to provide an indication of the relative spread of the data around the mean is the coefficient of variation, CV = $[SD / \bar{x}]$ 100%.

Common Sources of Error

There are three common sources of error in any analytical technique:

— *Personal Errors (Blunders)*: These occur when the analytical test is not carried out correctly: the wrong chemical reagent or equipment might have been used; some of the sample may have been spilt; a volume or mass may have been recorded incorrectly; etc. It is partly for this reason that analytical measurements should be repeated a number of times using freshly prepared laboratory samples. Blunders are usually easy to identify and can be eliminated by carrying out the analytical method again more carefully.

— *Random Errors:* These produce data that vary in a non-reproducible fashion from one measurement to the next e.g., instrumental noise. This type of error determines the standard deviation of a measurement. There may be a number of different sources of random error and these are accumulative.

— *Systematic Errors:* A systematic error produces results that consistently deviate from the true answer in some systematic way; e.g., measurements may always be 10% too high. This type of error would occur if the volume of a pipette was different from the stipulated value. For example, a nominally 100 cm^3 pipette may always deliver 101 cm^3 instead of the correct value.

To make accurate and precise measurements it is important when designing and setting up an analytical procedure to identify the various sources of error and to minimise their effects. Often, one particular step will be the largest source of error, and the best improvement in accuracy or precision can be achieved by minimising the error in this step.

Propagation of Errors

Most analytical procedures involve a number of steps, and there will be an error associated with each step. These individual errors accumulate to determine the overall error in the final result. For random errors there are a number of simple rules that can be followed to calculate the error in the final result:

Addition (Z = X+Y) and Subtraction (Z = X-Y):

$$\Delta Z = \sqrt{\Delta X^2 + \Delta Y^2} \tag{3}$$

Multiplication (Z = XY) and Division (Z = X/Y):

$$\frac{\Delta Z}{Z} = \sqrt{[\frac{\Delta X}{X}]^2 + [\frac{\Delta Y}{Y}]^2} \tag{4}$$

Here, ΔX is the standard deviation of the mean value X, ΔY is the standard deviation of the mean value Y, and ΔZ is the standard deviation of the mean value Z. These simple rules should be learnt and used when calculating the overall error in a final result. As an example, let us assume that we want to determine the fat content of a food and that we have previously measured the mass of extracted fat extracted from the food (M_E) and the initial mass of the food (M_I):

$$M_E = 3.1\ 0.3\ g$$
$$M_I = 10.5\ 0.7\ g$$
$$\%\ Fat\ Content = 100\ M_E\ /\ M_I$$

To calculate the mean and standard deviation of the fat content we need to use the multiplication rule (Z=X/Y) given by Equation 4. Initially, we assign values to the various parameters in the appropriate propagation of error equation:

$$X = 3.1;\ \Delta X = 0.3$$
$$Y = 10.5;\ \Delta Y = 0.7$$
$$\%\ Fat\ Content = Z = 100\ X/Y = 100\ 3.1/10.5 = 29.5\%$$
$$\Delta Z = Z\ [(\Delta X/X)^2+(\Delta Y/Y)^2] = 29.5\%\ [(0.3/3.1)^2+(0.7/10.5)^2] = 3.5\%$$

Hence, the fat content of the food is 29.5 3.5%. In reality, it may be necessary to carry out a number of different steps in a calculation, some that involve addition/subtraction and some that involve multiplication/division. When carrying out multiplication/division calculations it is necessary to ensure that all appropriate addition/subtraction calculations have been completed first.

Significant Figures and Rounding

The number of significant figures used in reporting a final result is determined by the standard deviation of the measurements. A final result is reported to the correct number of significant figures when it contains all the digits that are known to be correct, plus a final one that is known to be uncertain. For example, a reported value of 12.13, means that the 12.1 is known to be correct but the 3 at the end is uncertain, it could be either a 2 or a 4 instead.

For multiplication ($Z = X \times Y$) and division ($Z = X/Y$), the significant figures in the final result (Z) should be equal to the significant figures in the number from which it was calculated (X or Y) that has the lowest significant figures.

When rounding numbers: always round any number with a final digit less than 5 downwards, and 5 or more upwards, e.g. 23.453 becomes 23.45; 23.455 becomes 23.46; 23.458 becomes 23.46. It is usually desirable to carry extra digits throughout the calculations and then round off the final result.

Standard Curves

When carrying out certain analytical procedures it is necessary to prepare standard curves that are used to determine some property of an unknown material. A series of calibration experiments is carried out using samples with known properties and a standard curve is plotted from this data. For example, a series of protein solutions with known concentration of protein could be prepared and their absorbance of electromagnetic radiation at 280 nm could be measured using a UV-visible spectrophotometer. For dilute protein solutions there is a linear relationship between absorbance and protein concentration:

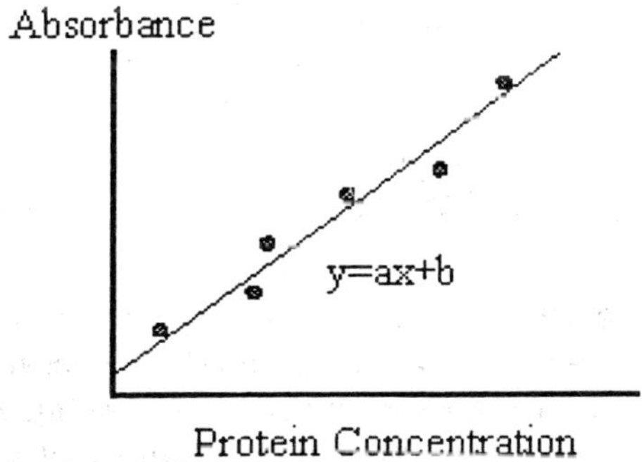

A best-fit line is drawn through the date using regression analysis, which has a gradient of *a* and a *y*-intercept of b. The concentration of protein in an unknown sample can then be determined by measuring its absorbance: $x = (y-b)/a$, where in this example x is the protein concentration and y is the absorbance.

How well the straight-line fits the experimental data is expressed by the correlation coefficient r^2, which has a value between 0 and 1.

The closer the value is to 1 the better the fit between the straight line and the experimental values: $r^2 = 1$ is a perfect fit. Most modern calculators and spreadsheet programmes have routines that can be used to automatically determine the regression coefficient, the slope and the intercept of a set of data.

Rejecting Data

When carrying out an experimental analytical procedure it will sometimes be observed that one of the measured values is very different from all of the other values, e.g., as the result of a "blunder" in the analytical procedure.

Occasionally, this value may be treated as being incorrect, and it can be rejected. There are certain rules based on statistics that allow us to decide whether a particular point can be rejected or not. A test called the *Q-test* is commonly used to decide whether an experimental value can be rejected or not.

$$Q = \frac{X_{BAD} - X_{NEXT}}{X_{HIGH} - X_{LOW}}$$

Here X_{BAD} is the questionable value, X_{NEXT} is the next closet value to X_{BAD}, X_{HIGH} is the highest value of the data set and X_{LOW} is the lowest value of the data set. If the Q-value is higher than the value given in a *Q-test* table for the number of samples being analysed then it can be rejected:

Number of Observations	Q-value for Data Rejection (90% confidence level)
3	0.94
4	0.76
5	0.64
6	0.56
7	0.51
8	0.47
9	0.44
10	0.41

For example, if five measurements were carried out and one measurement was very different from the rest (e.g., 20,22,25,50,21), having a *Q*-value of 0.84, then it could be safely rejected (because it is higher than the value of 0.64 given in the *Q*-test table for five observations).

References

Amigo L., *et al.*, "Genetic polymorphism of ovine milk proteins: its influence on technological properties of milk" – a review, *International Daairy Journal*, 10: 135–149, 2000.

Bania J., *et al.*, "Application of polymerase chain reaction for detection of goats' milk adulteration by milk of cow", *Journal of Dairy Research*, 68: 333–336, 2001.

Cattaneo T.M.P., *et al.*, "Analysis of cow, goat and ewe milk mixtures by capillary zone electrophoresis (CZE): preliminary approach", *Milchwissenschaft*, 51: 616–619, 1996.

Covello VT, *Merkhofer MW Risk assessment methods: approaches for assessing health and environmental risks*, Plenum Press, New York, 1993.

Morgan MG, *Henrion M Uncertainty: A guide to dealing with uncertainty in quantitative risk and policy analysis*, Cambridge University Press, Cambridge, 1990.

Kramer, A. and Twigg, B.A., *Fundamentals of quality control for the food industry*, AVI Publishing Co., Westport, Conn. USA, 1962.

Science and Judgment in Risk Assessment, *Committee on Risk Assessment of Hazardous Air Pollutants, Board on Environmental Studies and Toxicology*, National Research Council (NRC), National Academy of Sciences, National Academy Press, Washington DC, 1994.

Nielsen, S.S., *Food Analysis*, 2nd Edition, Aspen Publication, Gaithersberg, Maryland, 1998.

Procter, A. and Meullenet, J.F., "Sampling and Sample Preparation", *In: Food Analysis*, 2nd Edition, Aspen Publication, Gaithersberg, Maryland, 1998.

4

Food Protein Analysis

Proteins are important constituents of foods for a number of different reasons. They are a major source of energy, as well as containing essential amino-acids, such as lysine, tryptophan, methionine, leucine, isoleucine and valine, which are essential to human health, but which the body cannot synthesize. Proteins are also the major structural components of many natural foods, often determining their overall texture.

Isolated proteins are often used in foods as ingredients because of their unique functional properties, i.e., their ability to provide desirable appearance, texture or stability. Typically, proteins are used as gelling agents, emulsifiers, foaming agents and thickeners.

Many food proteins are enzymes which are capable of enhancing the rate of certain biochemical reactions. These reactions can have either a favourable or detrimental effect on the overall properties of foods. Food analysts are interested in knowing the total concentration, type, molecular structure and functional properties of the proteins in foods.

Determination of Protein Concentration

Kjeldahl Method

The Kjeldahl method was developed in 1883 by a brewer called Johann Kjeldahl. A food is digested with a strong acid so that it releases nitrogen which can be determined by a suitable titration technique. The amount of protein present is then calculated from the nitrogen concentration of the food. The same basic approach is still used today, although a number of improvements have been made to speed up the process and to obtain more accurate measurements.

It is usually considered to be the standard method of determining protein concentration. Because the Kjeldahl method does not measure the protein content directly a conversion factor (F) is needed to convert the measured nitrogen concentration to a protein concentration. A conversion factor of 6.25 (equivalent to 0.16 g nitrogen per gram of protein) is used for many applications, however, this is only an average value, and each protein has a different conversion factor depending on its amino-acid composition. The Kjeldahl method can conveniently be divided into three steps: digestion, neutralisation and titration.

Principles

Digestion

The food sample to be analysed is weighed into a digestion flask and then digested by heating it in the presence of sulfuric acid (an oxidizing agent which digests the food), anhydrous sodium sulfate (to speed up the reaction by raising the boiling point) and a catalyst, such as copper, selenium, titanium, or mercury (to speed up the reaction).

Digestion converts any nitrogen in the food (other than that which is in the form of nitrates or nitrites) into ammonia, and other organic matter to CO_2 and H_2O. Ammonia gas is not liberated in an acid solution because the ammonia is in the form of the ammonium ion (NH_4^+) which binds to the sulfate ion (SO_4^{2-}) and thus remains in solution:

$$N_{(food)} \rightarrow (NH_4)_2SO_4 \qquad (1)$$

Neutralisation

After the digestion has been completed the digestion flask is connected to a receiving flask by a tube. The solution in the digestion flask is then made alkaline by addition of sodium hydroxide, which converts the ammonium sulfate into ammonia gas:

$$(NH_4)_2SO_4 + 2\ NaOH \rightarrow 2NH_3 + 2H_2O + Na_2SO_4 \qquad (2)$$

The ammonia gas that is formed is liberated from the solution and moves out of the digestion flask and into the receiving flask—which contains an excess of boric acid. The low pH of the solution in the receiving flask converts the ammonia gas into the ammonium ion, and simultaneously converts the boric acid to the borate ion:

$$NH_3 + H_3BO_3 \text{ (boric acid)} \rightarrow NH_4^+ + H_2BO_3^- \text{ (borate ion)} \qquad (3)$$

Titration of the ammonium borate

The nitrogen content is then estimated by titration of the ammonium borate formed with standard sulfuric or hydrochloric acid, using a suitable indicator to determine the end-point of the reaction.

$$H_2BO_3^- + H^+ \rightarrow H_3BO_3 \qquad (4)$$

The concentration of hydrogen ions (in moles) required to reach the end-point is equivalent to the concentration of nitrogen that was in the original food (Equation 3). The following equation can be used to determine the nitrogen concentration of a sample that weighs m grams using a xM HCl acid solution for the titration:

$$\%\,N = \frac{x\,moles}{1000\,cm^3} \times \frac{(v_s - v_b)\,cm^3}{m\,g} \times \frac{14\,g}{moles} \times 100 \qquad (5)$$

Where v_s and v_b are the titration volumes of the sample and blank, and 14g is the molecular weight of nitrogen N. A blank sample is usually ran at the same time as the material being analysed to take into account any residual nitrogen which may be in the reagents used to carry out the analysis. Once the nitrogen content has been determined it is converted to a protein content using the appropriate conversion factor: %Protein = F %N.

Advantages and Disadvantages

Advantages: The Kjeldahl method is widely used internationally and is still the standard method for comparison against all other methods. Its universality, high precision and good reproducibility have made it the major method for the estimation of protein in foods.

Disadvantages: It does not give a measure of the true protein, since all nitrogen in foods is not in the form of protein. Different proteins need different correction factors because they have different amino acid sequences. The use of concentrated sulfuric acid at high temperatures poses a considerable hazard, as does the use of some of the possible catalysts The technique is time consuming to carry-out.

Dumas Method

Recently, an automated instrumental technique has been developed which is capable of rapidly measuring the protein concentration of food samples. This technique is based on a method first described by a scientist called Dumas over a century and a half ago. It is beginning to compete with the Kjeldahl method as the standard method of analysis for proteins for some foodstuffs due to its rapidness.

Principles

A sample of known mass is combusted in a high temperature (about 900°C) chamber in the presence of oxygen. This leads to the release of CO_2, H_2O and N_2. The CO_2 and H_2O are removed by passing the gasses over special columns that absorb them. The nitrogen content is then measured by passing the remaining gasses through a column that has a thermal conductivity detector at the end. The column helps separate the nitrogen from any residual CO_2 and H_2O that may have remained in the gas stream.

The instrument is calibrated by analysing a material that is pure and has a known nitrogen concentration, such as EDTA

(= 9.59%N). Thus the signal from the thermal conductivity detector can be converted into a nitrogen content. As with the Kjeldahl method it is necessary to convert the concentration of nitrogen in a sample to the protein content, using suitable conversion factors which depend on the precise amino acid sequence of the protein.

Advantages and Disadvantages

Advantages: It is much faster than the Kjeldahl method (under 4 minutes per measurement, compared to 1-2 hours for Kjeldahl). It doesn't need toxic chemicals or catalysts. Many samples can be measured automatically. It is easy to use.

Disadvantages: High initial cost. It does not give a measure of the true protein, since all nitrogen in foods is not in the form of protein. Different proteins need different correction factors because they have different amino acid sequences. The small sample size makes it difficult to obtain a representative sample.

UV-visible Spectroscopy Methods

A number of methods have been devised to measure protein concentration, which are based on UV-visible spectroscopy. These methods use either the natural ability of proteins to absorb (or scatter) light in the UV-visible region of the electromagnetic spectrum, or they chemically or physically modify proteins to make them absorb (or scatter) light in this region. The basic principle behind each of these tests is similar. First of all a calibration curve of absorbance (or turbidity) versus protein concentration is prepared using a series of protein solutions of known concentration.

The absorbance (or turbidity) of the solution being analysed is then measured at the same wavelength, and its protein concentration determined from the calibration curve. The main difference between the tests are the chemical groups which are responsible for the absorption or scattering of radiation, e.g., peptide bonds, aromatic side-groups, basic groups and aggregated proteins. A number of the most commonly used UV-visible methods for determining the protein content of foods are highlighted below:

Principles

Direct measurement at 280nm

Tryptophan and tyrosine absorb ultraviolet light strongly at 280 nm. The tryptophan and tyrosine content of many proteins remains fairly constant, and so the absorbance of protein solutions at 280nm can be used to determine their concentration. The advantages of this method are that the procedure is simple to carry out, it is nondestructive, and no special reagents are required.

The major disadvantage is that nucleic acids also absorb strongly at 280 nm and could therefore interfere with the measurement of the protein if they are present in sufficient concentrations. Even so, methods have been developed to overcome this problem, e.g., by measuring the absorbance at two different wavelengths.

Biuret Method

A violet-purplish colour is produced when cupric ions (Cu^{2+}) interact with peptide bonds under alkaline conditions. The biuret reagent, which contains all the chemicals required to carry out the analysis, can be purchased commercially. It is mixed with a protein solution and then allowed to stand for 15-30 minutes before the absorbance is read at 540 nm.

The major advantage of this technique is that there is no interference from materials that adsorb at lower wavelengths, and the technique is less sensitive to protein type because it utilises absorption involving peptide bonds that are common to all proteins, rather than specific side groups. However, it has a relatively low sensitivity compared to other UV-visible methods.

Lowry Method

The Lowry method combines the biuret reagent with another reagent (the Folin-Ciocalteau phenol reagent) which reacts with tyrosine and tryptophan residues in proteins. This gives a bluish colour which can be read somewhere between 500 - 750 nm depending on the sensitivity required. There is a small peak around 500 nm that can be used to determine high protein concentrations and a large peak around 750 nm that can be used to determine low protein concentrations. This method is more sensitive to low concentrations of proteins than the biuret method.

Dye Binding Methods

A known excess of a negatively charged (anionic) dye is added to a protein solution whose pH is adjusted so that the proteins are positively charged. The proteins form an insoluble complex with the dye because of the electrostatic attraction between the molecules, but the unbound dye remains soluble.

The anionic dye binds to cationic groups of the basic amino acid residues (histidine, arganine and lysine) and to free amino terminal groups. The amount of unbound dye remaining in solution after the insoluble protein-dye complex has been removed is determined by measuring its absorbance. The amount of protein present in the original solution is proportional to the amount of dye that bound to it: dyebound = dyeinitial - dyefree.

Turbimetric Method

Protein molecules which are normally soluble in solution can be made to precipitate by the addition of certain chemicals, e.g., trichloroacetic acid. Protein precipitation causes the solution to become turbid. Thus the concentration of protein can be determined by measuring the degree of turbidity.

Advantages and Disadvantages

Advantages: UV-visible techniques are fairly rapid and simple to carry out, and are sensitive to low concentrations of proteins.

Disadvantages: For most UV-visible techniques it is necessary to use dilute and transparent solutions, which contain no contaminating substances which absorb or scatter light at the same wavelength as the protein being analysed. The need for transparent solutions means that most foods must undergo significant amounts of sample preparation before they can be analysed, e.g., homogenisation, solvent extraction, centrifugation, filtration, which can be time consuming and laborious.

In addition, it is sometimes difficult to quantitatively extract proteins from certain types of foods, especially after they have been processed so that the proteins become aggregated or covalently bound with other substances. In addition the absorbance depends on the type of protein analysed (different proteins have different amino acid sequences).

Instrumental Methods

There are a wide variety of different instrumental methods available for determining the total protein content of food materials. These can be divided into three different categories according to their physicochemical principles:

(i) measurement of bulk physical properties,

(ii) measurement of adsorption of radiation, and

(iii) measurement of scattering of radiation.

Each instrumental methods has its own advantages and disadvantages, and range of foods to which it can be applied.

Principles

Measurement of bulk physical properties

— *Density*: The density of a protein is greater than that of most other food components, and so there is an increase in density of a food as its protein content increases. Thus the protein content of foods can be determined by measuring their density.

— *Refractive index*: The refractive index of an aqueous solution increases as the protein concentration increases and therefore RI measurements can be used to determine the protein content.

Measurement of adsorption of radiation

— *UV-visible*: The concentration of proteins can be determined by measuring the absorbance of ultraviolet-visible radiation.

— *Infrared*: Infrared techniques can be used to determine the concentration of proteins in food samples. Proteins absorb IR naturally due to characteristic vibrations (stretching and bending) of certain chemical groups along the polypeptide backbone. Measurements of the absorbance of radiation at certain wavelengths can thus be used to quantify the concentration of protein in the sample. IR is particularly useful for rapid on-line analysis of protein content. It also requires little sample preparation and is nondestructive. Its major disadvantages are its high initial cost and the need for extensive calibration.

— *Nuclear Magnetic Resonance*: NMR spectroscopy can be used to determine the total protein concentration of foods. The protein content is determined by measuring the area under a peak in an NMR chemical shift spectra that corresponds to the protein fraction.

Measurement of scattering of radiation

— *Light scattering*: The concentration of protein aggregates in aqueous solution can be determined using light scattering techniques because the turbidity of a solution is directly proportional to the concentration of aggregates present.

— *Ultrasonic scattering*: The concentration of protein aggregates can also be determined using ultrasonic scattering techniques because the ultrasonic velocity and absorption of ultrasound are related to the concentration of protein aggregates present.

Advantages and Disadvantages

A number of these instrumental methods have major advantages over the other techniques mentioned above because they are nondestructive, require little or no sample preparation, and measurements are rapid and precise. A major disadvantage of the techniques which rely on measurements of the bulk physical properties of foods are that a calibration curve must be prepared between the physical property of interest and the total protein content, and this may depend on the type of protein present and the food matrix it is contained within.

In addition, the techniques based on measurements of bulk physicochemical properties can only be used to analyse foods with relatively simple compositions. In a food that contains many different components whose concentration may vary, it is difficult to disentangle the contribution that the protein makes to the overall measurement from that of the other components.

Comparison of Methods

As food scientists we may often be in a position where we have to choose a particular technique for measuring the protein concentration of a food. How do we decide which technique is the most appropriate for our particular application? The first thing to determine is what is the information going to be used for. If the analysis is to be carried out for official purposes, e.g., legal or labeling requirements, then it is important to use an officially recognised method.

The Kjeldahl method, and increasingly the Dumas method, have been officially approved for a wide range of food applications. In contrast, only a small number of applications of UV-visible spectroscopy have been officially recognised. For quality control purposes, it is often more useful to have rapid and simple measurements of protein content and therefore IR techniques are most suitable.

For fundamental studies in the laboratory, where pure proteins are often analysed, UV-visible spectroscopic techniques are often preferred because they give rapid and reliable measurements, and are sensitive to low concentrations of protein. Other factors which may have to be considered are the amount of sample preparation required, their sensitivity and their speed.

The Kjeldahl, Dumas and IR methods require very little sample preparation. After a representative sample of the food has been selected it can usually be tested directly. On the other hand, the various UV-visible methods require extensive sample preparation prior to analysis. The protein must be extracted from the food into a dilute transparent solution, which usually involves time consuming homogenisation, solvent extraction, filtration and centrifugation procedures. In addition, it may be difficult to completely isolate some proteins from foods because they are strongly bound to other components.

The various techniques also have different sensitivities, i.e., the lowest concentration of protein which they can detect. The UV-visible methods are the most sensitive, being able to detect protein concentrations as low as 0.001 wt%. The sensitivity of the Dumas, Kjeldahl and IR methods is somewhere around 0.1 wt%. The time required per analysis, and the number of samples which can be run simultaneously, are also important factors to consider when deciding which analytical technique to use. IR techniques are capable of rapid analysis (< 1 minute) of protein concentration once they have been calibrated. The modern instrumental Dumas method is fully automated and can measure the protein concentration of a sample in less than 5 minutes, compared to the Kjeldahl method which takes between 30 minutes and 2 hours to carry out.

The various UV-visible methods range between a couple of minutes to an hour (depending on the type of dye that is used and how long it takes to react), although it does have the advantage that many samples can be run simultaneously. Nevertheless, it is usually necessary to carry out extensive sample preparation prior to analysis in order to get a transparent solution. Other factors which may be important when selecting an appropriate technique are: the equipment available, ease of operation, the desired accuracy, and whether or not the technique is nondestructive.

Characterisation of Protein Separation

Food analysts are often interested in the type of proteins present in a food because each protein has unique nutritional and physicochemical properties. Protein type is usually determined by separating and isolating the individual proteins from a complex mixture of proteins, so that they can be subsequently identified and characterised. Proteins are separated on the basis of differences in their physicochemical properties, such as size, charge, adsorption characteristics, solubility and heat-stability.

The choice of an appropriate separation technique depends on a number of factors, including the reasons for carrying out the analysis, the amount of sample available, the desired purity, the equipment available, the type of proteins present and the cost. Large-scale methods are available for crude isolations of large quantities of proteins, whereas small-scale methods are available for proteins that are expensive or only available in small quantities.

One of the factors that must be considered during the separation procedure is the possibility that the native three dimensional structure of the protein molecules may be altered. A prior knowledge of the effects of environmental conditions on protein structure and interactions is extremely useful when selecting the most appropriate separation technique. Firstly, because it helps determine the most suitable conditions to use to isolate a particular protein from a mixture of proteins, and secondly, because it may be important to choose conditions which will not adversely affect the molecular structure of the proteins.

Separation Techniques

Proteins can be separated by exploiting differences in their solubility in aqueous solutions. The solubility of a protein molecule is determined by its amino acid sequence because this determines its size, shape, hydrophobicity and electrical charge. Proteins can be selectively precipitated or solubilized by altering the pH, ionic strength, dielectric constant or temperature of a solution. These separation techniques are the most simple to use when large quantities of sample are involved, because they are relatively quick, inexpensive and are not particularly influenced by other food components. They are often used as the first step in any separation procedure because the majority of the contaminating materials can be easily removed.

Salting Out

Proteins are precipitated from aqueous solutions when the salt concentration exceeds a critical level, which is known as salting-out, because all the water is "bound" to the salts, and is therefore not available to hydrate the proteins. Ammonium sulfate [$(NH_4)_2SO_4$] is commonly used because it has a high water-solubility, although other neutral salts may also be used, e.g., NaCl or KCl. Generally a two-step procedure is used to maximise the separation efficiency. In the first step, the salt is added at a concentration just below that necessary to precipitate out the protein of interest.

The solution is then centrifuged to remove any proteins that are less soluble than the protein of interest. The salt concentration is then increased to a point just above that required to cause precipitation of the protein. This precipitates out the protein of interest (which can be separated by centrifugation), but leaves more soluble proteins in solution. The main problem with this method is that large concentrations of salt contaminate the solution, which must be removed before the protein can be resolubilsed.

Isoelectric Precipitation

The isoelectric point (pI) of a protein is the pH where the net charge on the protein is zero. Proteins tend to aggregate and precipitate at their pI because there is no electrostatic repulsion keeping them apart. Proteins have different isoelectric points because of their different amino acid sequences, and thus they can be separated by adjusting the pH of a solution. When the pH is adjusted to the pI of a particular protein it precipitates leaving the other proteins in solution.

Solvent Fractionation

The solubility of a protein depends on the dielectric constant of the solution that surrounds it because this alters the magnitude of the electrostatic interactions between charged groups. As the dielectric constant of a solution decreases the magnitude of the electrostatic interactions between charged species increases. This tends to decrease the solubility of proteins in solution because they are less ionized, and therefore the electrostatic repulsion between them is not sufficient to prevent them from aggregating.

The dielectric constant of aqueous solutions can be lowered by adding water-soluble organic solvents, such as ethanol or acetone. The amount of organic solvent required to cause precipitation depends on the protein and therefore proteins can be separated on this basis. The optimum quantity of organic solvent required to precipitate a protein varies from about 5 to 60%. Solvent fractionation is usually performed at 0°C or below to prevent protein denaturation caused by temperature increases that occur when organic solvents are mixed with water.

Denaturation

Many proteins are denatured and precipitate from solution when heated above a certain temperature or by adjusting a solution to highly acid or basic pHs. Proteins that are stable at high temperature or at extremes of pH are most easily separated by this technique because contaminating proteins can be precipitated while the protein of interest remains in solution.

Adsorption Chromatography

Adsorption chromatography involves the separation of compounds by selective adsorption-desorption at a solid matrix that is contained within a column through which the mixture passes. Separation is based on the different affinities of different proteins for the solid matrix. Affinity and ion-exchange chromatography are the two major types of adsorption chromatography commonly used for the separation of proteins. Separation can be carried out using either an open column or high-pressure liquid chromatography.

Ion Exchange Chromatography

Ion exchange chromatography relies on the reversible adsorption-desorption of ions in solution to a charged solid matrix or polymer network. This technique is the most commonly used chromatographic technique for protein separation. A positively charged matrix is called an anion-exchanger because it binds negatively charged ions (anions). A negatively charged matrix is called a cation-exchanger because it binds positively charged ions (cations).

The buffer conditions (pH and ionic strength) are adjusted to favour maximum binding of the protein of interest to the ion-exchange column. Contaminating proteins bind less strongly and therefore pass more rapidly through the column. The protein of interest is then eluted using another buffer solution which favors its desorption from the column.

Affinity Chromatography

Affinity chromatography uses a stationary phase that consists of a ligand covalently bound to a solid support. The ligand is a molecule that has a highly specific and unique reversible affinity for a particular protein. The sample to be analysed is passed through the column and the protein of interest binds to the ligand, whereas the contaminating proteins pass directly through.

The protein of interest is then eluted using a buffer solution which favors its desorption from the column. This technique is the most efficient means of separating an individual protein from a mixture of proteins, but it is the most expensive, because of the need to have columns with specific ligands bound to them. Both ion-exchange and affinity chromatography are commonly used to separate proteins and amino-acids in the laboratory. They are used less commonly for commercial separations because they are not suitable for rapidly separating large volumes and are relatively expensive.

Separation Due to Size Differences

Proteins can also be separated according to their size. Typically, the molecular weights of proteins vary from about 10,000 to 1,000,000 daltons. In practice, separation depends on the Stokes radius of a protein, rather than directly on its molecular weight. The Stokes radius is the average radius that a protein has in solution, and depends on its three dimensional molecular structure. For proteins with the same molecular weight the Stokes radius increases in the following order: compact globular protein < flexible random-coil < rod-like protein.

Dialysis

Dialysis is used to separate molecules in solution by use of semipermeable membranes that permit the passage of molecules smaller than a certain size through, but prevent the passing of larger molecules. A protein solution is placed in dialysis tubing which is sealed and placed into a large volume of water or buffer which is slowly stirred. Low molecular weight solutes flow through the bag, but the large molecular weight protein molecules remain in the bag. Dialysis is a relatively slow method, taking up to 12 hours to be completed. It is therefore most frequently used in the laboratory. Dialysis is often used to remove salt from protein solutions after they have been separated by salting-out, and to change buffers.

Ultrafiltration

A solution of protein is placed in a cell containing a semipermeable membrane, and pressure is applied. Smaller molecules pass through the membrane, whereas the larger molecules remain in the solution. The separation principle of this technique is therefore similar to dialysis, but because pressure is applied separation is much quicker. Semipermeable membranes with cutoff points between about 500 to 300,000 are available. That portion of the solution which is retained by the cell (large molecules) is called the retentate, whilst that part which passes through the membrane (small molecules) forms part of the ultrafiltrate. Ultrafiltration can be used to concentrate a protein solution, remove salts, exchange buffers or fractionate proteins on the basis of their size. Ultrafiltration units are used in the laboratory and on a commercial scale.

Size Exclusion Chromatography

This technique, sometimes known as gel filtration, also separates proteins according to their size. A protein solution is poured into a column which is packed with porous beads made of a cross-linked polymeric material. Molecules larger than the pores in the beads are excluded, and move quickly through the column, whereas the movement of molecules which enter the pores is retarded. Thus molecules are eluted off the column in order of decreasing size. Beads of different average pore size are available for separating proteins of different molecular weights.

Manufacturers of these beads provide information about the molecular weight range that they are most suitable for separating. Molecular weights of unknown proteins can be determined by comparing their elution volumes Vo, with those determined using proteins of known molecular weight: a plot of elution volume versus log(molecular weight) should give a straight line. One problem with this method is that the molecular weight is not directly related to the Stokes radius for different shaped proteins.

Electrophoresis

Electrophoresis relies on differences in the migration of charged molecules in a solution when an electrical field is applied across it. It can be used to separate proteins on the basis of their size, shape or charge.

Non-denaturing Electrophoresis

In non-denaturing electrophoresis, a buffered solution of native proteins is poured onto a porous gel and a voltage is applied across the gel. The proteins move through the gel in a direction that depends on the sign of their charge, and at a rate that depends on the magnitude of the charge, and the friction to their movement:

$$mobility = \frac{applied\ voltage \times molecular\ charge}{molecular\ friction}$$

Proteins may be positively or negatively charged in solution depending on their isoelectic points (pI) and the pH of the solution. A protein is negatively charged if the pH is above the pI, and positively charged if the pH is below the pI. The magnitude of the charge and applied voltage will determine how far proteins migrate in a certain time. The higher the voltage or the greater the charge on the protein the further it will move.

The friction of a molecule is a measure of its resistance to movement through the gel and is largely determined by the relationship between the effective size of the molecule, and the size of the pores in the gel. The smaller the size of the molecule, or the larger the size of the pores in the gel, the lower the resistance and therefore the faster a molecule moves through the gel.

Gels with different porosity's can be purchased from chemical suppliers, or made up in the laboratory. Smaller pores sizes are obtained by using a higher concentration of cross-linking reagent to form the gel. Gels may be contained between two parallel plates, or in cylindrical tubes. In non-denaturing electrophoresis the native proteins are separated based on a combination of their charge, size and shape.

Denaturing Electrophoresis Proteins

In denaturing electrophoresis proteins are separated primarily on their molecular weight. Proteins are denatured prior to analysis by mixing them with mercaptoethanol, which breaks down disulfide bonds, and sodium dodecyl sulfate (SDS), which is an anionic surfactant that hydrophobically binds to protein molecules and causes them to unfold because of the repulsion between negatively charged surfactant head-groups. Each protein molecule binds approximately the same amount of SDS per unit length. Hence, the charge per unit length and the molecular conformation is approximately similar for all proteins.

As proteins travel through a gel network they are primarily separated on the basis of their molecular weight because their movement depends on the size of the protein molecule relative to the size of the pores in the gel: smaller proteins moving more rapidly through the matrix than larger molecules. This type of electrophoresis is commonly called sodium dodecyl sulfate -polyacrylamide gel electrophoresis, or SDS-PAGE.

To determine how far proteins have moved a tracking dye is added to the protein solution, e.g., bromophenol blue. This dye is a small charged molecule that migrates ahead of the proteins. After the electrophoresis is completed the proteins are made visible by treating the gel with a protein dye such as Coomassie Brilliant Blue or silver stain. The relative mobility of each protein band is calculated:

$$R_m = \frac{distance\ protein\ moves}{distance\ dye\ moves}$$

Electrophoresis is often used to determine the protein composition of food products. The protein is extracted from the food into solution, which is then separated using electrophoresis.

SDS-PAGE is used to determine the molecular weight of a protein by measuring R_m, and then comparing it with a calibration curve produced using proteins of known molecular weight: a plot of log (molecular weight) against relative mobility is usually linear. Denaturing electrophoresis is more useful for determining molecular weights than non-denaturing electrophoresis, because the friction to movement does not depend on the shape or original charge of the protein molecules.

Isoelectric Focusing Electrophoresis

This technique is a modification of electrophoresis, in which proteins are separated by charge on a gel matrix which has a pH gradient across it. Proteins migrate to the location where the pH equals their isoelectric point and then stop moving because they are no longer charged. This methods has one of the highest resolutions of all techniques used to separate proteins. Gels are available that cover a narrow pH range (2-3 units) or a broad pH range (3-10 units) and one should therefore select a gel which is most suitable for the proteins being separated.

Two Dimensional Electrophoresis

Isoelectric focusing and SDS-Page can be used together to improve resolution of complex protein mixtures. Proteins are separated in one direction on the basis of charge using isoelectric focusing, and then in a perpendicular direction on the basis of size using SDS-Page.

Amino Acid Analysis

Amino acid analysis is used to determine the amino acid composition of proteins. A protein sample is first hydrolyzed to release the amino acids, which are then separated using chromatography.

References

Cox G., "Famine Symposium—the Ecology of Famine: An Overview," *Ecology of Food and Nutrition*, 1978, 6:207-220.

Francks P., "Agriculture and State in Industrial East Asia: The Rise and Fall of the Food Control System in Japan," *Japan Forum*, 1998, 10:1.

Hopfenberg R., "Human Carrying Capacity Is Determined by Food Availability," *Population and Environment*, 2003, 25:109-117(9).

Lee R., "Monogongo: The Ethnography of a Major Wild Food Resource," *Ecology of Food and Nutrition*, 1973, 3:89-105.

5

Analysis of Lipid Concentration in Food

In many foods the lipid component plays a major role in determining the overall physical characteristics, such as flavour, texture, mouthfeel and appearance. For this reason, it is difficult to develop low-fat alternatives of many foods, because once the fat is removed some of the most important physical characteristics are lost. Finally, many fats are prone to lipid oxidation, which leads to the formation of off-flavours and potentially harmful products. Some of the most important properties of concern to the food analyst are:

- Total lipid concentration
- Type of lipids present
- Physicochemical properties of lipids, e.g., crystallisation, melting point, smoke point, rheology, density and colour
- Structural organisation of lipids within a food

Properties of Lipids

Lipids are usually defined as those components that are soluble in organic solvents (such as ether, hexane or chloroform), but are insoluble in water. This group of substances includes triacylglycercols, diacylglycercols, monoacylglycercols, free fatty acids, phospholipids, sterols, caretonoids and vitamins A and D. The lipid fraction of a fatty food therefore contains a complex mixture of different types of molecule.

Even so, triacylglycercols are the major component of most foods, typically making up more than 95 to 99% of the total lipids present. Triacylglycerols are esters of three fatty acids and a glycerol molecule. The fatty acids normally found in foods vary in chain length, degree of unsaturation and position on the glycerol molecule. Consequently, the triacylglycerol fraction itself consists of a complex mixture of different types of molecules.

Each type of fat has a different profile of lipids present which determines the precise nature of its nutritional and physiochemical properties. The terms fat, oil and lipid are often used interchangeably by food scientists. Although sometimes the term fat is used to describe those lipids that are solid at the specified temperature, whereas the term oil is used to describe those lipids that are liquid at the specified temperature.

Sample Preparation Procedure

As with any analytical procedure, the validity of the results depends on proper sampling and preservation of the sample prior to analysis. Ideally, the composition of the sample analysed should represent as closely as possible that of the food from which it was taken. The sample preparation required in lipid analysis depends on the type of food being analysed, the nature of the lipid component and the type of analytical procedure used.

In order, to decide the most appropriate sample preparation procedure it is necessary to have a knowledge of the physical structure and location of the principal lipids present in the food. Since each food is different it is necessary to use different procedures for each one. Official methods have been developed for specific types of foods that stipulate the precise sample preparation procedure that should be followed.

In general, sample preparation should be carried out using an environment that minimises any changes in the properties of the lipid fraction. If lipid oxidation is a problem it is important to preserve the sample by using a nitrogen atmosphere, cold temperature, low light or adding antioxidants. If the solid fat content or crystal structure is important it may be necessary to carefully control the temperature and handling of the sample.

Determination of Lipid Concentration

It is important to be able to accurately determine the total fat content of foods for a number of reasons:

— Economic (not to give away expensive ingredients)

— Legal (to conform to standards of identity and nutritional labelling laws)

— Health (development of low fat foods)

— Quality (food properties depend on the total lipid content)

— Processing (processing conditions depend on the total lipid content)

The principle physicochemical characteristics of lipids (the "analyte") used to distinguish them from the other components in foods (the "matrix") are their solubility in organic solvents, immiscibility with water, physical characteristics and spectroscopic properties. The analytical techniques based on these principles can be conveniently categorised into three different types:

(i) solvent extraction;

(ii) non-solvent extraction and

(iii) instrumental methods.

Solvent Extraction Techniques

The fact that lipids are soluble in organic solvents, but insoluble in water, provides the food analyst with a convenient method of separating the lipid components in foods from water soluble components, such as proteins, carbohydrates and minerals. In fact, solvent extraction techniques are one of the most commonly used methods of isolating lipids from foods and of determining the total lipid content of foods.

Sample Preparation for Solvent Extraction

The preparation of a sample for solvent extraction usually involves a number of steps:

- *Drying sample*: It is often necessary to dry samples prior to solvent extraction, because many organic solvents cannot easily penetrate into foods containing water, and therefore extraction would be inefficient.
- *Particle size reduction*: Dried samples are usually finely ground prior to solvent extraction to produce a more homogeneous sample and to increase the surface area of lipid exposed to the solvent. Grinding is often carried out at low temperatures to reduce the tendency for lipid oxidation to occur.
- *Acid hydrolysis*: Some foods contain lipids that are complexed with proteins (lipoproteins) or polysaccharides (glycolipids). To determine the concentration of these components it is necessary to break the bonds which hold the lipid and non-lipid components together prior to solvent extraction. Acid hydrolysis is commonly used to release bound lipids into easily extractable forms, e.g. a sample is digested by heating it for 1 hour in the presence of 3N HCl acid.
- *Solvent Selection*: The ideal solvent for lipid extraction would completely extract all the lipid components from a food, while leaving all the other components behind. In practice, the efficiency of solvent extraction depends on the polarity of the lipids present compared to the polarity of the solvent. Polar lipids (such as glycolipids or phospholipids) are more soluble in polar solvents (such as alcohols), than in non-polar solvents (such as hexane).

On the other hand, non-polar lipids (such as triacylglycerols) are more soluble in non-polar solvents than in polar ones. The fact that different lipids have different polarities means that it is impossible to select a single organic solvent to extract them all. Thus the total lipid content determined by solvent extraction depends on the nature of the organic solvent used to carry out the extraction: the total lipid content determined using one solvent may be different from that determined using another solvent.

In addition to the above considerations, a solvent should also be inexpensive, have a relatively low boiling point (so that it can easily be removed by evaporation), be non-toxic and be nonflammable (for safety reasons). It is difficult to find a single solvent which meets all of these requirements. Ethyl ether and petroleum ether are the most commonly used solvents, but pentane and hexane are also used for some foods.

Batch Extraction Process

These methods are based on mixing the sample and the solvent in a suitable container, e.g., a separatory funnel. The container is shaken vigorously and the organic solvent and aqueous phase are allowed to separate (either by gravity or centrifugation). The aqueous phase is then decanted off, and the concentration of lipid in the solvent is determined by evaporating the solvent and measuring the mass of lipid remaining: %Lipid = 100 (M_{lipid}/M_{sample}). This procedure may have to be repeated a number of times to improve the efficiency of the extraction process.

In this case the aqueous phase would undergo further extractions using fresh solvent, then all the solvent fractions would be collected together and the lipid determined by weighing after evaporation of solvent. The efficiency of the extraction of a particular type of lipid by a particular type of solvent can be quantified by an equilibrium partition coefficient, $K = c_{solvent}/c_{aqueous}$, where csolvent and caqueous are the concentration of lipid in the solvent and aqueous phase, respectively. The higher the partition coefficient the more efficient the extraction process.

Semi-Continuous Solvent Extraction Methods

Semi-continuous solvent extraction methods are commonly used to increase the efficiency of lipid extraction from foods. The Soxhlet method is the most commonly used example of a semi-continuous method. In the Soxhlet method a sample is dried, ground into small particles and placed in a porous thimble. The thimble is placed in an extraction chamber, which is suspended above a flask containing the solvent and below a condenser.

The flask is heated and the solvent evaporates and moves up into the condenser where it is converted into a liquid that trickles into the extraction chamber containing the sample. Eventually, the solvent builds up in the extraction chamber and completely surrounds the sample. The extraction chamber is designed so that when the solvent surrounding the sample exceeds a certain level it overflows and trickles back down into the boiling flask. As the solvent passes through the sample it extracts the lipids and carries them into the flask.

The lipids then remain in the flask because of their low volatility. At the end of the extraction process, which typically lasts a few hours, the flask containing the solvent and lipid is removed, the solvent is evaporated and the mass of lipid remaining is measured (Mlipid). The percentage of lipid in the initial sample (Msample) can then be calculated: %Lipid = 100 (Mlipid/Msample). A number of instrument manufacturers have designed modified versions of the Soxhlet method that can be used to determine the total lipid content more easily and rapidly.

Continuous Solvent Extraction

The Goldfish method is similar to the Soxhlet method except that the extraction chamber is designed so that the solvent just trickles through the sample rather than building up around it. This reduces the amount of time required to carry out the extraction, but it has the disadvantage that channelling of the solvent can occur, i.e., the solvent may preferentially take certain routes through the sample and therefore the extraction is inefficient. This is not a problem in the Soxhlet method because the sample is always surrounded by solvent.

Accelerated Solvent Extraction

The efficiency of solvent extraction can be increased by carrying it out at a higher temperature and pressure than are normally used. The effectiveness of a solvent at extracting lipids from a food increases as its temperature increases, but the pressure must also be increased to keep the solvent in the liquid state. This reduces the amount of solvent required to carry out the analysis, which is beneficial from a cost and environmental standpoint. Special instruments are available to carry out solvent extraction at elevated temperatures and pressures.

Supercritical Fluid Extraction

Solvent extraction can be carried out using special instruments that use supercritical carbon dioxide (rather than organic liquids) as the solvent. These instruments are finding greater use because of the cost and environmental problems associated with the usage and disposal of organic solvents. When pressurised CO_2 is heated above a certain critical temperature it becomes a supercritical fluid, which has some of the properties of a gas and some of a liquid. The fact that it behaves like a gas means that it can easily penetrate into a sample and extract the lipids, while the fact that it behaves like a fluid means that it can dissolve a large quantity of lipids (especially at higher pressures).

Instruments based on this principle heat the food sample to be analysed in a pressurised chamber and then mix supercritical CO_2 fluid with it. The CO_2 extracts the lipid, and forms a separate solvent layer, which is separated from the aqueous components. The pressure and temperature of the solvent are then reduced which causes the CO_2 to turn to a gas, leaving the lipid fraction remaining. The lipid content of a food is determined by weighing the percentage of lipid extracted from the original sample.

Nonsolvent Liquid Extraction

A number of liquid extraction methods do not rely on organic solvents, but use other chemicals to separate the lipids from the rest of the food. The Babcock, Gerber and Detergent methods are examples of nonsolvent liquid extraction methods for determining the lipid content of milk and some other dairy products.

Babcock Method

A specified amount of milk is accurately pipetted into a specially designed flask (the Babcock bottle). Sulfuric acid is mixed with the milk, which digests the protein, generates heat, and breaks down the fat globule membrane that surrounds the droplets, thereby releasing the fat. The sample is then centrifuged while it is hot (55-60°C) which causes the liquid fat to rise into the neck of the Babcock bottle. The neck is graduated to give the amount of milk fat present in wt%. The Babcock method takes about 45 minutes to carry out, and is precise to within 0.1%. It does not determine phospholipids in milk, because they are located in the aqueous phase or at the boundary between the lipid and aqueous phases.

Gerber Method

This method is similar to the Babcock method except that a mixture of sulfuric acid and isoamyl alcohol, and a slightly different shaped bottle, are used. It is faster and simpler to carry out than the Babcock method. The isoamyl alcohol is used to prevent charring of the sugars by heat and sulfuric acid which can be a problem in the Babcock method since it makes it difficult to read the fat content from the graduated flask. This method is used mainly in Europe, whilst the Babcock method is used mainly in the USA. As with the Babcock method, it does not determine phospholipids.

Detergent Method

This method was developed to overcome the inconvenience and safety concerns associated with the use of highly corrosive acids. A sample is mixed with a combination of surfactants in a Babcock bottle. The surfactants displace the fat globule membrane which surrounds the emulsion droplets in milk and causes them to coalesce and separate. The sample is centrifuged which allows the fat to move into the graduated neck of the bottle, where its concentration can then be determined.

Instrumental Methods

There are a wide variety of different instrumental methods available for determining the total lipid content of food materials. These can be divided into three different categories according to their physicochemical principles:

(i) measurement of bulk physical properties,

(ii) measurement of adsorption of radiation, and

(iii) measurement of scattering of radiation.

Each instrumental methods has its own advantages and disadvantages, and range of foods to which it can be applied.

Measurement of Bulk Physical Properties

— *Density*: The density of liquid oil is less than that of most other food components, and so there is a decrease in density of a food as its fat content increases. Thus the lipid content of foods can be determined by measuring their density.

— *Electrical conductivity*: The electrical conductivity of lipids is much smaller than that of aqueous substances, and so the conductivity of a food decreases as the lipid concentration increases. Measurements of the overall electrical conductivity of foods can therefore be used to determine fat contents.

— *Ultrasonic velocity*: The speed at which an ultrasonic wave travels through a material depends on the concentration of fat in a food. Thus the lipid content can be determined by measuring its ultrasonic velocity. This technique is capable of rapid, nondestructive on-line measurements of lipid content.

Measurement of Adsorption of Radiation

— *UV-visible*: The concentration of certain lipids can be determined by measuring the absorbance of ultraviolet-visible radiation. The lipid must usually be extracted and diluted in a suitable solvent prior to analysis, thus the technique can be quite time-consuming and labour intensive.

- *Infrared*: This method is based on the absorbance of IR energy at a wavelength of 5.73 m due to molecular vibrations or rotations associated with fat molecules: the greater the absorbance the more fat present. IR is particularly useful for rapid and on-line analysis of lipid content once a suitable calibration curve has been developed.
- *Nuclear Magnetic Resonance*: NMR spectroscopy is routinely used to determine the total lipid concentration of foods. The lipid content is determined by measuring the area under a peak in an NMR chemical shift spectra that corresponds to the lipid fraction. Lipid contents can often be determined in a few seconds without the need for any sample preparation using commercially available instruments.
- *X-ray absorption*: Lean meat absorbs X-rays more strongly than fat, thus the X-ray absorbance decreases as the lipid concentration increases. Commercial instruments have been developed which utilise this phenomenon to determine the lipid content of meat and meat products.

Measurement of Scattering of Radiation

- *Light scattering*: The concentration of oil droplets in dilute food emulsions can be determined using light scattering techniques because the turbidity of an emulsion is directly proportional to the concentration of oil droplets present.
- *Ultrasonic scattering*: The concentration of oil droplets in concentrated food emulsions can be determined using ultrasonic scattering techniques because the ultrasonic velocity and absorption of ultrasound by an emulsion is related to the concentration of oil droplets present.

A number of these instrumental methods have major advantages over the extraction techniques mentioned above because they are nondestructive, require little or no sample preparation, and measurements are usually rapid, precise and simple. A major disadvantage of the techniques which rely on measurements of the bulk physical properties of foods are that a calibration curve must be prepared between the physical property of interest and the total lipid content, and this may depend on the type of lipid present and the food matrix it is contained in.

In addition, these techniques can only be used to analyse foods with relatively simple compositions. In a food that contains many different components whose concentration may vary, it is difficult to disentangle the contribution that the fat makes to the overall measurement from that of the other components.

Comparison of Methods

Soxhlet extraction is one of the most commonly used methods for determination of total lipids in dried foods. This is mainly because it is fairly simple to use and is the officially recognised method for a wide range of fat content determinations. The main disadvantages of the technique are that a relatively dry sample is needed (to allow the solvent to penetrate), it is destructive, and it is time consuming. For high moisture content foods it is often better to use batch solvent or nonsolvent extraction techniques.

Many instrumental methods are simple to operate, rapid, reproducible, require little sample preparation and are nondestructive. Nevertheless, they are often expensive to purchase and can only be used for certain types of foods, i.e., where there is no interference from other components. In addition, calibration curves prepared for instrumental methods usually require that the fat content be measured using a standard method.

Extraction techniques tend to be more accurate and more generally applicable and are therefore the standard methods for official analysis of many food materials. Instrumental methods are most useful for rapid measurements of fat content on-line or in quality assurance laboratories of food factories where many samples must be measured rapidly.

Lipid Composition

Lipids are an extremely diverse group of compounds consisting of tri-, di- and monoacylglycercols, free fatty acids, phospholipids, sterols, caretonoids and vitamins A and D. In addition, most of these sub-groups are themselves chemically complex. All triacylglycerols are esters of glycerol and three fatty acid molecules, nevertheless, the fatty acids can have different chain lengths, branching, unsaturation, and positions on the glycerol molecule.

Thus even a lipid which consists of only triacylglycerols may contain a huge number of different chemical species. It is often important for food scientists to either know or to be able to specify the concentration of the different types of lipid molecules present, as well as the total lipid concentration. Some of the most important reasons for determining the type of lipids present in foods are listed below:

— *Legal*: Government regulations often demand that the amounts of saturated, unsaturated and polyunsaturated lipids, as well as the amount of cholesterol, be specified on food labels.

— *Food Quality*: Desirable physical characteristics of foods, such as appearance, flavour, mouthfeel and texture, depend on the type of lipids present.

— *Lipid oxidation*: Foods which contain high concentrations of unsaturated lipids are particularly susceptible to lipid oxidation, which can lead to the formation of undesirable off-flavours and aromas, as well as potentially toxic compounds e.g., cholesterol oxides.

— *Adulteration*: Adulteration of fats and oils can be detected by measuring the type of lipids present, and comparing them with the profile expected for an unadulterated sample.

— *Food Processing*: The manufacture of many foods relies on a knowledge of the type of lipids present in order to adjust the processing conditions to their optimum values, e.g. temperatures, flow rates etc.

Sample Chosen for Analysis

It is important that the sample chosen for analysis is representative of the lipids present in the original food, and that its properties are not altered prior to the analysis. Analysis of the types of lipids present in a food usually requires that the lipid be available in a fairly pure form. Thus foods which are almost entirely lipids, such as olive oil, vegetable oil or lard, can usually be analysed with little sample preparation. Nevertheless, for many other foods it is necessary to extract and purify the lipid component prior to analysis. Lipids can sometimes be extracted by simply applying pressure to a food to squeese out the oil, e.g., some fish, nuts and seeds.

For most foods, however, more rigorous extraction methods are needed, such as the solvent or nonsolvent extraction methods described in the previous lecture. Once the lipids have been separated they are often melted and then filtered or centrifuged to remove any extraneous matter. In addition, they are often dried to remove any residual moisture which might interfere with the analysis. As with any analytical procedure it is important not to alter the properties of the component being analysed during the extraction process. Oxidation of unsaturated lipids can be minimised by adding antioxidants, or by flushing containers with nitrogen gas and avoiding exposure to heat and light.

Chromatography

Chromatography is one of the most powerful analytical procedures for separating and analysing the properties of lipids, especially when combined with techniques which can be used to identify the chemical structure of the peaks, e.g., mass spectrometry or NMR. A chromatographic analysis involves passing a mixture of the molecules to be separated through a column that contains a matrix capable of selectively retarding the flow of the molecules.

Molecules in the mixture are separated because of their differing affinities for the matrix in the column. The stronger the affinity between a specific molecule and the matrix, the more its movement is retarded, and the slower it passes through the column. Thus different molecules can be separated on the basis of the strength of their interaction with the matrix. After being separated by the column, the concentration of each of the molecules is determined as they pass by a suitable detector.

Chromatography can be used to determine the complete profile of molecules present in a lipid. This information can be used to: calculate the amounts of saturated, unsaturated, polyunsaturated fat and cholesterol; the degree of lipid oxidation; the extent of heat or radiation damage; detect adulteration; determine the presence of antioxidants. Various forms of chromatography are available to analyse the lipids in foods, e.g. thin layer chromatography (TLC), gas chromatography (GC), and high pressure liquid chromatography (HPLC).

Lipid Fractions by TLC

TLC is used mainly to separate and determine the concentration of different types of lipid groups in foods, e.g. triacylglycerols, diacylglycerols, monoacylglycerols, cholesterol, cholesterol oxides and phospholipids. A TLC plate is coated with a suitable absorbing material and placed into an appropriate solvent. A small amount of the lipid sample to be analysed is spotted onto the TLC plate. With time the solvent moves up the plate due to capillary forces and separates different lipid fractions on the basis of their affinity for the absorbing material.

At the end of the separation the plate is sprayed with a dye so as to make the spots visible. By comparing the distance that the spots move with standards of known composition it is possible to identify the lipids present. Spots can be scraped off and analysed further using techniques, such as GC, NMR or mass spectrometry. This procedure is inexpensive and allows rapid analysis of lipids in fatty foods.

Fatty Acid Methyl Esters by GC

Intact triacylglycerols and free fatty acids are not very volatile and are therefore difficult to analyse using GC (which requires that the lipids be capable of being volatised in the instrument). For this reason lipids are usually derivitised prior to analysis to increase their volatility. Triacylglycerols are first saponified which breaks them down to glycerol and free fatty acids, and are then methylated.

$$\text{Triacylglycerol} \xrightarrow{CH_1OH,N_2OH} \text{Fatty acid methyl esters (FAMEs) +}$$
$$\text{methylated glycerol}$$

Saponification reduces the molecular weight and methylation reduces the polarity, both of which increase the volatility of the lipids. The concentration of different volatile fatty acid methyl esters (FAMEs) present in the sample is then analysed using GC. The FAMES are dissolved in a suitable organic solvent that is then injected into a GC injection chamber.

The sample is heated in the injection chamber to volatilise the FAMES and then carried into the separating column by a heated carrier gas. As the FAMES pass through the column they are separated into a number of peaks based on differences in their molecular weights and polarities, which are quantified using a suitable detector. Determination of the total fatty acid profile allows one to calculate the type and concentration of fatty acids present in the original lipid sample

Chemical Techniques

A number of chemical methods have been developed to provide information about the type of lipids present in edible fats and oils. These techniques are much cruder than chromatography techniques, because they only give information about the average properties of the lipid components present, e.g. the average molecular weight, degree of unsaturation or amount of acids present. Nevertheless, they are simple to perform and do not require expensive apparatus, and so they are widely used in industry and research.

Iodine Value (IV)

The iodine value (IV) gives a measure of the average degree of unsaturation of a lipid: the higher the iodine value, the greater the number of C=C double bonds. By definition the iodine value is expressed as the grams of iodine absorbed per 100g of lipid. One of the most commonly used methods for determining the iodine value of lipids is "Wijs method". The lipid to be analysed is weighed and dissolved in a suitable organic solvent, to which a known excess of iodine chloride is added. Some of the ICl reacts with the double bonds in the unsaturated lipids, while the rest remains:

$$R\text{-}CH{=}CH\text{-}R + ICl_{excess} \rightarrow R\text{-}CHI\text{-}CHCl\text{-}R + ICl_{remaining}$$

The amount of ICl that has reacted is determined by measuring the amount of $ICl_{remaining}$ after the reaction has gone to completion ($ICl_{reacted} = ICl_{excess} - ICl_{remaining}$). The amount of $ICl_{remaining}$ is determined by adding excess potassium iodide to the solution to liberate iodine, and then titrating with a sodium thiosulfate ($Na_2S_2O_3$) solution in the presence of starch to determine the concentration of iodine released:

$$ICl_{remaining} + 2KI \rightarrow KCl + KI + I_2$$

$$I_2 + \text{starch} + 2Na_2S_2O_3\ (\text{blue}) \rightarrow 2NaI + \text{starch} + Na_2S_4O_6\ (\text{colourless})$$

Iodine itself has a reddish brown colour, but this is often not intense enough to be used as a good indication of the end-point of the reaction. For this reason, starch is usually used as an indicator because it forms a molecular complex with the iodine that has a deep blue colour. Initially, starch is added to the solution that contains the iodine and the solution goes a dark blue. Then, the solution is titrated with a sodium thiosulfate solution of known molarity.

While there is any I_2 remaining in the solution it stays blue, but once all of the I_2 has been converted to I it turns colourless.

Thus, a change in solution appearance from blue to colourless can be used as the end-point of the titration. The concentration of C=C in the original sample can therefore be calculated by measuring the amount of sodium thiosulfate needed to complete the titration. The higher the degree of unsaturation, the more iodine absorbed, and the higher the iodine value. The iodine value is used to obtain a measure of the average degree of unsaturation of oils, and to follow processes such as hydrogenation and oxidation that involve changes in the degree of unsaturation.

Saponification Number

The saponification number is a measure of the average molecular weight of the triacylglycerols in a sample. Saponification is the process of breaking down a neutral fat into glycerol and fatty acids by treatment with alkali:

Triacylglycerol + 3 KOH $\rightarrow$ Glycerol + 3 Fatty acid salts of potassium

The saponification number is defined as the mg of KOH required to saponify one gram of fat. The lipid is first extracted and then dissolved in an ethanol solution which contains a known excess of KOH. This solution is then heated so that the reaction goes to completion. The unreacted KOH is then determined by adding an indicator and titrating the sample with HCl. The saponification number is then calculated from a knowledge of the weight of sample and the amount of KOH which reacted. The smaller the saponification number the larger the average molecular weight of the triacylglycerols present.

Acid Value

The acid value is a measure of the amount of free acids present in a given amount of fat. The lipids are extracted from the food sample and then dissolved in an ethanol solution containing an indicator. This solution is then titrated with alkali (KOH) until a pinkish colour appears. The acid value is defined as the mg of KOH necessary to neutralise the fatty acids present in 1g of lipid. The acid value may be overestimated if other acid components are present in the system, e.g. amino acids or acid phosphates. The acid value is often a good measure of the break down of the triacylglycrols into free fatty acids, which has an adverse effect on the quality of many lipids.

Instrumental Methods

A variety of instrumental methods can also be used to provide information about lipid composition. The most powerful of these is nuclear magnetic resonance (NMR) spectroscopy. By measuring the chemical shift spectra it is possible to determine the concentration of specific types of chemical groups present, which can be used to estimate the concentration of different types of lipids. Indirect information about the average molecular weight and degree of unsaturation of the oils can be obtained by measuring physical properties, such as density or refractive index.

The refractive index increases with increasing chain length and increasing unsaturation, whereas the density decreases with increasing chain length and decreasing unsaturation. Measurements of the refractive index or density can therefore be used to monitor processes that involve a change in the composition of oils, e.g. hydrogenation, which decreases the degree of unsaturation.

Analysing Lipid Oxidation in Foods

Foods which contain high concentrations of unsaturated lipids are particularly susceptible to lipid oxidation. Lipid oxidation is one of the major forms of spoilage in foods, because it leads to the formation of off-flavours and potentially toxic compounds. Lipid oxidation is an extremely complex process involving numerous reactions that give rise to a variety of chemical and physical changes in lipids:

reactants → primary products → secondary products

(unsaturated lipids and O_2) → (peroxides and conjugated dienes) → (ketones,aldehydes,alcohols,hydrocarbons)

Food scientists have developed a number of methods to characterise the extent of lipid oxidation in foods, and to determine whether or not a particular lipid is susceptible to oxidation.

Chromatography

Chromatography is the most powerful method of monitoring lipid oxidation because it provides a detailed profile of the fatty acids and other molecules present in lipids. Valuable information about the lipid oxidation process is obtained by measuring changes in this profile with time, especially when peaks are identified using mass spectrometry or NMR. It is possible to monitor the loss of reactants and the formation of specific reaction products using chromatography. These measurements may be made on non-polar lipids extracted from the food, water-soluble reaction products present in the aqueous phase of a food or volatile components in the head-space of a food.

Lipid Oxidation

Lipid oxidation depends on the reaction between unsaturated fatty acids and oxygen. Thus it is possible to monitor the rate at which it occurs by measuring the uptake of oxygen by the sample as the reaction proceeds. Usually, the lipid is placed in a sealed container and the amount of oxygen that must be input into the container to keep the oxygen concentration in the head-space above the sample constant is measured. The more oxygen that has to be fed into the container, the faster the rate of lipid oxidation. This technique is therefore an example of a measurement of the reduction in the concentration of reactants.

Peroxides

Peroxides (R-OOH) are primary reaction products formed in the initial stages of oxidation, and therefore give an indication of the progress of lipid oxidation. One of the most commonly used methods to determine peroxide value utilises the ability of peroxides to liberate iodine from potassium iodide. The lipid is dissolved in a suitable organic solvent and an excess of KI is added:

$$ROOH + KI_{excess} \rightarrow ROH + KOH + I_2$$

Once the reaction has gone to completion, the amount of ROOH that has reacted can be determined by measuring the amount of iodine formed. This is done by titration with sodium thiosulfate and a starch indicator:

$$I_2 + \text{starch} + 2Na_2S_2O_3 \text{ (blue)} \rightarrow 2NaI + \text{starch} + Na_2S_4O_6 \text{ (colourless)}$$

The amount of sodium thiosulfate required to titrate the reaction is related to the concentration of peroxides in the original sample. There are a number of problems with the use of peroxide value as an indication of lipid oxidation. Firstly, peroxides are primary products that are broken down in the latter stages of lipid oxidation.

Thus, a low value of PV may represent either the initial or final stages of oxidation. Secondly, the results of the procedure are highly sensitive to the conditions used to carry out the experiment, and so the test must always be standardised. This technique is an example of a measurement of the increase in concentration of primary reaction products.

Conjugated Dienes

Almost immediately after peroxides are formed, the non-conjugated double bonds (C=C-C-C=C) that are present in natural unsaturated lipids are converted to conjugated double bonds (C=C-C=C). Conjugated dienes absorb ultraviolet radiation strongly at 233nm, whereas conjugated trienes absorb at 268nm. Thus oxidation can be followed by dissolving the lipid in a suitable organic solvent and measuring the change in its absorbance with time using a UV-visible spectrophotometer.

In the later stages of lipid oxidation the conjugated dienes (which are primary products) are broken down into secondary products (which do not adsorb UV-visible light strongly) which leads to a decrease in absorbance. This method is therefore only useful for monitoring the early stages of lipid oxidation. This technique is an example of a measurement of the increase in concentration of primary reaction products.

Thiobarbituric Acid (TBA)

This is one of the most widely used tests for determining the extent of lipid oxidation. It measures the concentration of relatively polar secondary reaction products, i.e., aldehydes. The lipid to be analysed is dissolved in a suitable non-polar solvent which is contained within a flask. An aqueous solution of TBA reagent is added to the flask and the sample is shaken, which causes the polar secondary products to be dissolved in it. After shaking the aqueous phase is separated from the non-polar solvent, placed in a test-tube, and heated for 20 minutes in boiling water, which produces a pink colour.

The intensity of this pink colour is directly related to the concentration of TBA-reactive substances in the original sample, and is determined by measuring its absorbance at 540 nm using a UV-visible spectrophotometer. The principle source of colour is the formation of a complex between TBA and malanoaldehyde, although some other secondary reaction products can also react with the TBA reagent. For this reason, this test is now usually referred to as the thiobarbituric acid reactive substances (TBARS) method. TBARS is an example of a measurement of the increase in concentration of secondary reaction products.

Lipid Oxidation Tests

Rather than determining the extent of lipid oxidation in a particular food, it is often more important to know its susceptibility to oxidation. Normally, oxidation can take a long time to occur, e.g., a few days to a few months, which is impractical for routine analysis. For this reason, a number of accelerated oxidation tests have been developed to speed up this process. These methods artificially increase the rate of lipid oxidation by exposing the lipid to heat, oxygen, metal catalysts, light or enzymes. Even so there is always some concern that the results of accelerated tests do not adequately model lipid oxidation in real systems.

A typical accelerated oxidation test is the active oxygen method (AOM). A liquid sample is held at 98°C while air is constantly bubbled through it. Stability is expressed as hours of heating until rancidity occurs, which may be determined by detection of a rancid odour or by measuring the peroxide value. Another widely used accelerated oxidation test is the Schaal Oven Test. A known weight of oil is placed in an oven at a specified temperature (about 65 °C) and the time until rancidity is detected is recorded by sensory evaluation or measuring the peroxide value.

Physicochemical Properties

In addition to their nutritional importance lipids are also used in foods because of their characteristic physicochemical properties, such as mouthfeel, flavour, texture and appearance. They are also used as heat transfer agents during the preparation of other foods, e.g. for frying. It is therefore important for food scientists to have analytical techniques that can be used to characterise the physicochemical properties of lipids.

Solid Fat Content (SFC)

The solid fat content (SFC) of a lipid influences many of its sensory and physical properties, such as spreadability, firmness, mouthfeel, processing and stability. Food manufacturers often measure the variation of SFC with temperature when characterising lipids that are used in certain foods, e.g., margarine and butter. The solid fat content is defined as the percentage of the total lipid that is solid at a particular temperature, i.e. $SFC = 100M_{solid}/M_{total}$, where M_{solid} is the mass of the lipid that is solid and M_{total} is the total mass of the lipid in the food.

A variety of methods have been developed to measure the temperature dependence of the solid fat content. The density of solid fat is higher than the density of liquid oil, and so there is an increase in density when a fat crystallises and a decrease when it melts. By measuring the density over a range of temperatures it is possible to determine the solid fat content—temperature profile:

$$SFC = \frac{(\rho - \rho_L)}{(\rho_S - \rho_L)} \times 100$$

where ρ is the density of the lipid at a particular temperature, and ρ_L and ρ_S are the densities of the lipid if it were completely liquid or completely solid at the same temperature. The density is usually measured by density bottles or dilatometry. More recently, instrumental methods based on nuclear magnetic resonance (NMR) have largely replaced density measurements, because measurements are quicker and simpler to carry out (although the instrumentation is considerably more expensive).

Basically, the sample is placed into an NMR instrument and a radio frequency pulse is applied to it. This induces a NMR signal in the sample, whose decay rate depends on whether the lipid is solid or liquid. The signal from the solid fat decays much more rapidly than the signal from the liquid oil and therefore it is possible to distinguish between these two components.

Techniques based on differential scanning calorimetry are also commonly used to monitor changes in SFC. These techniques measure the heat evolved or absorbed by a lipid when it crystallises or melts. By making these measurements over a range of temperatures it is possible to determine the melting point, the total amount of lipid involved in the transition and the SFC-temperature profile.

Melting Point

In many situations, it is not necessary to know the SFC over the whole temperature range, instead, only information about the temperature at which melting starts or ends is required. A pure triacylglycerol has a single melting point that occurs at a specific temperature. Nevertheless, foods lipids contain a wide variety of different triacylglycerols, each with their own unique melting point, and so they melt over a wide range of temperatures. Thus the "melting point" of a food lipid can be defined in a number of different ways, each corresponding to a different amount of solid fat remaining. Some of the most commonly used "melting points" are:

- *Clear point*: A small amount of fat is placed in a capillary tube and heated at a controlled rate. The temperature at which the fat completely melts and becomes transparent is called the "clear point".
- *Slip point*: A small amount of fat is placed in a capillary tube and heated at a controlled rate. The temperature at which the fat just starts to move downwards due to its weight is called the "slip point".
- *Wiley melting point*: A disc of fat is suspended in an alcohol-water mixture of similar density and is then heated at a controlled rate. The temperature at which the disc changes shape to a sphere is called the "Wiley melting point".

Cloud Point

This gives a measure of the temperature at which crystallisation begins in a liquid oil. A fat sample is heated to a temperature where all the crystals are known to have melted. The sample is then cooled at a controlled rate and the temperature at which the liquid just goes cloudy is determined. This temperature is known as the cloud point, and is the temperature where crystals begin to form and scatter light. It is often of practical importance to have an oil which does not crystallise when stored at 0°C for prolonged periods. A simple test to determine the ability of lipids to withstand cold temperatures without forming crystals, is to ascertain whether or not a sample goes cloudy when stored for 5 hours at 0°C.

Smoke, Flash and Fire Points

These tests give a measure of the effect of heating on the

physicochemical properties of lipids. They are particularly important for selecting lipids that are going to be used at high temperatures, e.g. during baking or frying. The tests reflect the amount of volatile organic material in oils and fats such as free fatty acids.

— The smoke point is the temperature at which the sample begins to smoke when tested under specified conditions. A fat is poured into a metal container and heated at a controlled rate in an oven. The smoke point is the temperature at which a thin continuous stream of bluish smoke is first observed.

— The flash point is the temperature at which a flash appears at any point on the surface of the sample due to the ignition of volatile gaseous products. The fat is poured into a metal container and heated at a controlled rate, with a flame being passed over the surface of the sample at regular intervals.

— The fire point is the temperature at which evolution of volatiles due to the thermal decomposition of the lipids proceeds so quickly that continuous combustion occurs (a fire).

Rheology of Lipids

The rheology of lipids is important in many food applications. Rheology is the science concerned with the deformation and flow of matter. Most rheological tests involve applying a force to a material and measuring its flow or change in shape. Many of the textural properties that people perceive when they consume foods are largely rheological in nature, e.g., creaminess, juiciness, smoothness, brittleness, tenderness, hardness, etc. The stability and appearance of foods often depends on the rheological characteristics of their components. The flow of foods through pipes or the ease at which they can be packed into containers are also determined by their rheology.

Liquid oils are usually characterised in terms of their flow properties (viscosity), whereas viscoelastic or plastic "solids" are characterised in terms of both their elastic (elastic modulus) and flow properties. A wide variety of experimental techniques are available to characterise the rheological properties of food materials. One of the most important rheological characteristics of lipids is their "plasticity", because this determines their "spreadability".

The plasticity of a lipid is due to the fact that fat crystals can form a three-dimensional network that gives the product some solid-like characteristics. Below a certain stress (known as the "yield stress") the product behaves like a solid with an elastic modulus because the crystal network is not disrupted, but above this stress it flows like a liquid because the crystal network is continually disrupted. Rheological techniques are therefore needed to measure the change in deformation of a lipid when stresses are applied.

References

Ackman, R.G., "Application of TLC to lipid separation: detection methods," In: *Analyses of Fats, Oils and Lipoproteins*, E.G. Perkins, American Oil Chemists' Society, Champaign, U.S.A. (ed.)., 1991, pp. 97-121.

Aitzetmuller, K., "The use of HPLC in modern fat analysis," *Fat Sci. Technol.*, 93, 501-510, 1991.

Christie,W.W., "Methods for stereospecific analysis of triacyl-sn-glycerols," In: *Contemporary Lipid Analysis, 2nd Symposium Proceedings*, N.U. Olsson & B.G. Herslof, LipidTeknik, Stockholm (ed.)., pp. 63-71, 1992.

Ebeler,S.E. and Shibamoto,T., "Overview and recent developments in solid-phase extraction for separation of lipid classes," In: *Lipid Chromatographic Analysis*, T. Shibamoto, Dekker, NY (ed.)., pp. 1-48, 1993.

Hetherington, A.M. and Drobak, B.K., "Inositol-containing lipids in higher plants," *Prog. Lipid Res.*, 31, 53-63, 1992.

Laker, M.F. and Game, F.L. "Approaches to lipid and lipoprotein analysis," *Baillieres Clin. Endocrin. Metab.*, 4, 693-717, 1990.

Patton,G.M. and Robins, S.J., "Extraction of phospholipids and analysis of phospholipid molecular species," *Methods Enzymol.*, 187, 195-215, 1990.

6

Determination of Ash and Minerals

Ash is the inorganic residue remaining after the water and organic matter have been removed by heating in the presence of oxidising agents, which provides a measure of the total amount of minerals within a food. Analytical techniques for providing information about the total mineral content are based on the fact that the minerals (the "analyte") can be distinguished from all the other components (the "matrix") within a food in some measurable way. The most widely used methods are based on the fact that minerals are not destroyed by heating, and that they have a low volatility compared to other food components.

The three main types of analytical procedure used to determine the ash content of foods are based on this principle: dry ashing, wet ashing and low temperature plasma dry ashing. The method chosen for a particular analysis depends on the reason for carrying out the analysis, the type of food analysed and the equipment available. Ashing may also be used as the first step in preparing samples for analysis of specific minerals, by atomic spectroscopy or the various traditional methods described below. Ash contents of fresh foods rarely exceed 5%, although some processed foods can have ash contents as high as 12%, e.g., dried beef.

Preparing Samples

As with all food analysis procedures it is crucial to carefully select a sample whose composition represents that of the food being analysed and to ensure that its composition does not change significantly prior to analysis. Typically, samples of 1-10g are used in the analysis of ash content. Solid foods are finely ground and then carefully mixed to facilitate the choice of a representative sample. Before carrying out an ash analysis, samples that are high in moisture are often dried to prevent spattering during ashing. High fat samples are usually defatted by solvent extraction, as this facilitates the release of the moisture and prevents spattering. Other possible problems include contamination of samples by minerals in grinders, glassware or crucibles which come into contact with the sample during the analysis. For the same reason, it is recommended to use deionised water when preparing samples.

Dry Ashing Procedures

Dry ashing procedures use a high temperature muffle furnace capable of maintaining temperatures of between 500 and 600 °C. Water and other volatile materials are vaporised and organic substances are burned in the presence of the oxygen in air to CO_2, H_2O and N_2. Most minerals are converted to oxides, sulphates, phosphates, chlorides or silicates. Although most minerals have fairly low volatility at these high temperatures, some are volatile and may be partially lost, e.g., iron, lead and mercury. If an analysis is being carried out to determine the concentration of one of these substances then it is advisable to use an alternative ashing method that uses lower temperatures.

The food sample is weighed before and after ashing to determine the concentration of ash present. The ash content can be expressed on either a dry or wet basis:

$$\%\ \text{Ash(dry basis)} = \frac{M_{ASH}}{M_{DRY}} \times 100$$

$$\%\ \text{Ash (wet basis)} = \frac{M_{ASH}}{M_{WET}} \times 100$$

where M_{ASH} refers to the mass of the ashed sample, and M_{DRY} and M_{ASH} refer to the original masses of the dried and wet samples.

There are a number of different types of crucible available for ashing food samples, including quartz, Pyrex, porcelain, steel and platinum. Selection of an appropriate crucible depends on the sample being analysed and the furnace temperature used. The most widely used crucibles are made from porcelain because it is relatively inexpensive to purchase, can be used up to high temperatures (< 1200°C) and are easy to clean.

Porcelain crucibles are resistant to acids but can be corroded by alkaline samples, and therefore different types of crucible should be used to analyse this type of sample. In addition, porcelain crucibles are prone to cracking if they experience rapid temperature changes. A number of dry ashing methods have been officially recognised for the determination of the ash content of various foods (AOAC Official Methods of Analysis). Typically, a sample is held at 500-600 oC for 24 hours.

Wet Ashing Procedures

Wet ashing is primarily used in the preparation of samples for subsequent analysis of specific minerals. It breaks down and removes the organic matrix surrounding the minerals so that they are left in an aqueous solution. A dried ground food sample is usually weighed into a flask containing strong acids and oxidising agents and then heated. Heating is continued until the organic matter is completely digested, leaving only the mineral oxides in solution. The temperature and time used depends on the type of acids and oxidising agents used. Typically, a digestion takes from 10 minutes to a few hours at temperatures of about 350°C. The resulting solution can then be analysed for specific minerals.

Low Temperature Plasma Ashing

A sample is placed into a glass chamber which is evacuated

using a vacuum pump. A small amount of oxygen is pumped into the chamber and broken down to nascent oxygen by application of an electromagnetic radio frequency field. The organic matter in the sample is rapidly oxidised by the nascent oxygen and the moisture is evaporated because of the elevated temperatures. The relatively cool temperatures (< 150°C) used in low-temperature plasma ashing cause less loss of volatile minerals than other methods.

Water Soluble and Insoluble Ash

As well as the total ash content, it is sometimes useful to determine the ratio of water soluble to water-insoluble ash as this gives a useful indication of the quality of certain foods, e.g., the fruit content of preserves and jellies. Ash is diluted with distilled water then heated to nearly boiling, and the resulting solution is filtered. The amount of soluble ash is determined by drying the filtrate, and the insoluble ash is determined by rinsing, drying and ashing the filter paper.

Comparison of Ashing Methods

The conventional dry ashing procedure is simple to carry out, is not labor intensive, requires no expensive chemicals and can be used to analyse many samples simultaneously. Nevertheless, the procedure is time-consuming and volatile minerals may be lost at the high temperatures used. Microwave instruments are capable of speeding up the process of dry ashing. Wet ashing and low temperature plasma ashing are more rapid and cause less loss of volatile minerals because samples are heated to lower temperatures. Nevertheless, the wet ashing procedure requires the use of hazardous chemicals and is labor intensive, while the plasma method requires expensive equipment and has a low sample throughput.

Specific Mineral Content

Knowledge of the concentration and type of specific minerals present in food products is often important in the food industry. The major physicochemical characteristics of minerals that are used to distinguish them from the surrounding matrix are: their low volatility; their ability to react with specific chemical reagents to give measurable changes; and their unique electromagnetic spectra.

The most effective means of determining the type and concentration of specific minerals in foods is to use atomic absorption or emission spectroscopy. Instruments based on this principle can be used to quantify the entire range of minerals in foods, often to concentrations as low as a few ppm. For these reasons they have largely replaced traditional methods of mineral analysis in institutions that can afford to purchase and maintain one, or that routinely analyse large numbers of samples. Institutions that do not have the resources or sample throughput to warrant purchasing an atomic spectroscopy instrument rely on more traditional methods that require chemicals and equipment commonly found in food laboratories. Many of the minerals of importance to food scientists can be measured using one of these traditional methods.

Sample Preparation

Many of the analytical methods used to determine the specific mineral content of foods require that the minerals be dissolved in an aqueous solution. For this reason, it is often necessary to isolate the minerals from the organic matrix surrounding them prior to the analysis. It is important that the ashing procedure does not alter the mineral concentration in the food due to volatilisation. Another potential source of error in mineral analysis is the presence of contaminants in the water, reagents or glassware. For this reason, ultrapure water or reagents should be used, and/or a blank should be run at the same time as the sample being analysed.

A blank uses the same glassware and reagents as the sample being analysed and therefore should contain the same concentration of any contaminants. The concentration of minerals in the blank is then subtracted from the value determined for the sample. Some substances can interfere with analysis of certain minerals, and should therefore be eliminated prior to the analysis or accounted for in the data interpretation. The principles of a number of the most important traditional methods for analysing minerals are described below. Many more traditional methods can be found in the AOAC Official Methods of Analysis.

Gravimetric Procedures

The element to be analysed is precipitated from solution by adding a reagent that reacts with it to form an insoluble complex with a known chemical formula. The precipitate is separated from the solution by filtration, rinsed, dried and weighed. The amount of mineral present in the original sample is determined from a knowledge of the chemical formula of the precipitate. For example, the amount of chloride in a solution can be determined by adding excess silver ions to form an insoluble silver chloride precipitate, because it is known that Cl is 24.74% of AgCl. Gravimetric procedures are only suitable for large food samples, which have relatively high concentrations of the mineral being analysed. They are not suitable for analysis of trace elements because balances are not sensitive enough to accurately weigh the small amount of precipitate formed.

Calorimetric Methods

These methods rely on a change in colour of a reagent when it reacts with a specific mineral in solution which can be quantified by measuring the absorbance of the solution at a specific wavelength using a spectrophotometer. Calorimetric methods are used to determine the concentration of a wide variety of different minerals. Vandate is often used as a calorimetric reagent because it changes colour when it reacts with minerals. For example, the phosphorous content of a sample can be determined by adding a vandate-molybdate reagent to the sample. This forms a coloured complex (yellow-orange) with the phosphorous which can be quantified by measuring the absorbance of the solution at 420nm, and comparing with a calibration curve. Different reagents are also available to colorimetrically determine the concentration of other minerals.

Effect of Titrations

EDTA Compleximetric Titration

EDTA is a chemical reagent that forms strong complexes with multivalent metallic ions. The disodium salt of EDTA is usually used because it is available in high purity: Na2H2Y. The complexes formed by metal ions and EDTA can be represented by the following equations:

$$m^{2+} + H_2Y^{2-} \rightarrow mY^{2}- + 2H^{+}$$

$$m^{3+} + H_2Y^{2-} \rightarrow mY- + 2H^{+}$$

$$m^{4+} + H_2Y^{2-} \rightarrow mY + 2H^{+}$$

The calcium content of foods is often determined by this method. An ashed food sample is diluted in water and then made alkaline (pH 12.5 to 13). An indicator that can form a coloured complex with EDTA is then added to the solution, and the solution is titrated with EDTA. The EDTA-indicator complex is chosen to be much weaker than the EDTA-mineral complex. Consequently, as long as multivalent ions remain in the solution the EDTA forms a strong complex with them and does not react with the indicator.

However, once all the mineral ions have been complexed, any additional EDTA reacts with the indicator and forms a coloured complex that is used to determine the end-point of the reaction. The calcium content of a food sample is determined by comparing the volume of EDTA required to titrate it to the end-point with a calibration curve prepared for a series of solutions of known calcium concentration.

If there is a mixture of different multivalent metallic ions present in a food there could be some problems in determining the concentration of a specific type of ion. It is often possible to remove interfering ions by passing the solution containing the sample through an ion-exchange column prior to analysis.

Redox Reactions

Many analytical procedures are based on coupled reduction-oxidation (redox) reactions. Reduction is the gain of electrons by atoms or molecules, whereas oxidation is the removal of electrons from atoms or molecules. Any molecular species that gains electrons during the course of a reaction is said to be reduced, whereas any molecular species that loses electrons is said to be oxidised, whether or not oxygen is involved. Electrons cannot be created or destroyed in ordinary chemical reactions and so any oxidation reaction is accompanied by a reduction reaction. These coupled reactions are called redox reactions:

$$X^n \rightarrow X^{n+1} + e^- \text{ (Oxidation reaction - loss of electrons)}$$

$$Y^m + e^- \rightarrow Y^{m-1} \text{ (Reduction reaction - gain of electrons)}$$

$$X^n + Y^m \rightarrow X^{n+1} + Y^{m-1} \text{ (Coupled reaction- transfer of electrons)}$$

Analysts often design a coupled reaction system so that one of the half-reactions leads to a measurable change in the system that can be conveniently used as an end-point, e.g., a colour change. Thus one of the coupled reactions usually involves the mineral being analysed (e.g., X = analyte), whereas the other involves an indicator (e.g., Y = indicator).

For example, permanganate ion (MnO_4^-) is a deep purple colour (oxidised form), while the mangenous ion (Mn^{2+}) is a pale pink colour (reduced form). Thus permanganate titrations can be used as an indicator of many redox reactions:

$$\underset{\text{(Deep Purple)}}{MnO_4^-} + 8H^+ + 5e^- \rightarrow \underset{\text{(Pale Pink)}}{Mn^{2+}} + 4H_20 \quad \text{(Reduction reaction)}$$

The calcium or iron content of foods can be determined by titration with a solution of potassium permanganate, the end point corresponding to the first change of the solution from pale pink to purple. The calcium or iron content is determined from the volume of permanganate solution of known molarity that is required to reach the end-point. For iron the reaction is:

$$5Fe^{2+} \rightarrow 5Fe^{3+} + 5e^{-} \text{ (Oxidation reaction)}$$

$$MnO_4^- + 8H^+ + 5e^- \rightarrow Mn^{2+} + 4H_20 \text{ (Reduction reaction)}$$

$$5Fe^{2+} + MnO_4^- + 8H^+ \rightarrow 5Fe^{3+} + Mn^{2+} + 4H_20 \text{ (Coupled reaction)}$$

Potassium permanganate is titrated into the aqueous solution of ashed food. While there is Fe^{2+} remaining in the food the MnO_4^- is converted to Mn^{2+} that leads to a pale pink solution. Once all of the Fe^{2+} has been converted to Fe^{3+} then the MnO_4^- remains in solution and leads to the formation of a purple colour, which is the end-point.

Precipitation Titrations

When at least one product of a titration reaction is an insoluble precipitate, it is referred to as a precipitation titration. A titrimetric method commonly used in the food industry is the Mohr method for chloride analysis. Silver nitrate is titrated into an aqueous solution containing the sample to be analysed and a chromate indicator.

$$AgNO_3 + NaCl \rightarrow AgCl(s) + NaNO_3$$

The interaction between silver and chloride is much stronger than that between silver and chromate. The silver ion therefore reacts with the chloride ion to form AgCl, until all of the chloride ion is exhausted. Any further addition of silver nitrate leads to the formation of silver chromate, which is an insoluble orange coloured solid.

$$Ag^+ + Cl^- \rightarrow AgCl \text{ (colourless) - until all Cl- is complexed}$$

$$2Ag^+ + CrO4^{2-} \rightarrow Ag_2CrO_4 \text{ (orange) - after all Cl- is complexed}$$

The end point of the reaction is the first hint of an orange colour. The volume of silver nitrate solution (of known molarity) required to reach the endpoint is determined, and thus the concentration of chloride in solution can be calculated.

Ion-Selective Electrodes

The mineral content of many foods can be determined using ion-selective electrodes (ISE). These devices work on the same principle as pH meters, but the composition of the glass electrode is different so that it is sensitive to specific types of ion (rather than H^+). Special glass electrodes are commercially available to determine the concentration of K^+, Na^+, NH_4^+, Li+, Ca^{2+} and Rb^+ in aqueous solution. Two electrodes are dipped into an aqueous solution containing the dissolved mineral: a reference electrode and a ion-selective electrode. The voltage across the electrodes depends on the concentration of the mineral in solution and is measured at extremely low current to prevent alterations in ion concentration.

The concentration of a specific mineral is determined from a calibration curve of voltage versus the logarithm of concentration. The major advantages of this method are its simplicity, speed and ease of use. The technique has been used to determine the salt concentration of butter, cheese and meat, the calcium concentration of milk and the CO_2 concentration of soft drinks. In principle, an ion selective electrode is only sensitive to one type of ion, however, there is often interference from other types of ions. This problem can often be reduced by adjusting pH, complexing or precipitating the interfering ions.

Finally, it should be noted that the ISE technique is only sensitive to the concentration of free ions present in a solution. If the ions are complexed with other components, such as chelating agents or biopolymers, then they will not be detected. The ISE technique is therefore particularly useful for quantifying the binding of minerals to food components. If one wants to determine the total concentration of a specific ion in a food (rather than the free concentration), then one needs to ensure that ion binding does not occur, e.g., by ashing the food.

Atomic Spectroscopy

The determination of mineral type and concentration by atomic spectroscopy is more sensitive, specific, and quicker than traditional wet chemistry methods. For this reason it has largely replaced traditional methods in laboratories that can afford it or that routinely analyse for minerals.

Principles of Atomic Spectroscopy

The primary cause of absorption and emission of radiation in atomic spectroscopy is electronic transitions of outer shell electrons. Photons with the energy associated with this type of transition are found in the UV-visible part of the electromagnetic spectrum. In this respect atomic spectroscopy is similar to UV-visible spectroscopy, however, the samples used in atomic spectroscopy are individual atoms in a gaseous state, whereas those used in UV-visible spectroscopy are molecules dissolved in liquids. This has important consequences for the nature of the spectra produced.

In atomic spectroscopy the peaks are narrow and well defined, but in UV-visible spectroscopy they are broad and overlap with one another. The are two major reasons for this. Firstly, because absorption or emission is from atoms, rather than molecules, there are no vibrational or rotational transitions superimposed on the electronic transitions. Secondly, because the atoms are in a gaseous state they are well separated from each other and do not interact with neighbouring molecules.

The energy change associated with a transition between two energy levels is related to the wavelength of the absorbed radiation: DE = hc/λ, where, h = Planks constant, c = the speed of light and the wavelength. Thus for a given transition between two energy states radiation of a discrete wavelength is either absorbed or emitted. Each element has a unique electronic structure and therefore it has a unique set of energy levels. Consequently, it absorbs or emits radiation at specific wavelengths. Each spectrum is therefore like a "fingerprint" that can be used to identify a particular element. In addition, because the absorption and emission of radiation occurs at different wavelengths for different types of atom, one element can be distinguished from others by making measurements at a wavelength where it absorbs or emits radiation, but the other elements do not.

Absorption occurs primarily when electrons in the ground state are promoted to various excited states. Emission occurs when electrons in an excited state fall back to a lower energy level. Atoms can exist in a number of different excited states, and can fall back to one of many different lower energy states (not necessarily the ground state). Thus there are many more lines in an emission spectra than there are in an absorption spectra.

Atomic spectroscopy is used to provide information about the type and concentration of minerals in foods. The type of minerals is determined by measuring the position of the peaks in the emission or absorption spectra. The concentration of mineral components is determined by measuring the intensity of a spectral line known to correspond to the particular element of interest.

The reduction in intensity of an electromagnetic wave that travels through a sample is used to determine the absorbance: $A = -\log(I/I_0)$. The Beer-Lambert law can then be used to relate the absorbance to the concentration of atoms in the sample: $A = a.b.c$, where A is absorbance, a is extinction coefficient, b is sample pathlength and c is concentration of absorbing species. In practice, there are often deviations from the above equation and so it is often necessary to prepare a calibration curve using a series of standards of known concentration prepared using the same reagents as used to prepare the sample. It is also important to run a blank to take into account any impurities in the reagents that might interfere with the analysis.

Atomic Absorption Spectroscopy

Atomic absorption spectroscopy (AAS) is an analytical

method that is based on the absorption of UV-visible radiation by free atoms in the gaseous state. The food sample to be analysed is normally ashed and then dissolved in an aqueous solution. This solution is placed in the instrument where it is heated to vaporise and atomise the minerals. A beam of radiation is passed through the atomised sample, and the absorption of radiation is measured at specific wavelengths corresponding to the mineral of interest. Information about the type and concentration of minerals present is obtained by measuring the location and intensity of the peaks in the absorption spectra.

Instrumentation

The radiation source: The most commonly used source of radiation in AAS is the hollow cathode lamp. This is a hollow tube filled with argon or neon, and a cathode filament made of the metallic form of the element to be analysed. When a voltage is applied across the electrodes, the lamp emits radiation characteristic of the metal in the cathode i.e., if the cathode is made of sodium, a sodium emission spectrum is produced. When this radiation passes through a sample containing sodium atoms it will be absorbed because it contains radiation of *exactly* the right wavelength to promote transition from one energy level to another. Thus a different lamp is needed for each type of element analysed.

Chopper: The radiation arriving at the detector comes from two different sources:

— radiation emitted by the filament of the lamp (which is partially absorbed by the sample);

— radiation that is emitted by the atoms in the sample that have been excited to higher energy levels by absorption of energy from the atomiser.

To quantify the concentration of minerals in a sample using AAS it is necessary to measure the reduction in amplitude of the beam of radiation that has passed through the sample, rather than the radiation emitted by the excited sample. This can be done using a mechanical device, called a chopper, in conjunction with an electronic device that distinguishes between direct and alternating currents.

The chopper is a spinning disk with a series of slits which is placed between the radiation source and the sample. The radiation from the light source is therefore continuously being switched on and off at a specific frequency, i.e., it is an alternating current. On the other hand, the radiation emitted from the excited atoms in the sample is constant i.e., it is direct current. The overall detected radiation is therefore the sum of a varying component and a constant component.

Electronic devices are available which can separate alternating and constant current. These devices are used in AAS instruments to isolate the signal generated by the light from that emitted by the atoms in the sample.

Atomiser: Atomisers are used to convert the sample to be analysed into individual atoms. The atomisation process is achieved by exposing the sample to high temperatures, and involves three stages:

— removal of water associated with molecules,

— conversion of molecules into a gas, and

— atomisation of molecules.

At higher temperatures the atoms may become ionised, which is undesirable because the atomic spectra of ionised atoms is different from that of non-ionised ones. Consequently, it is important to use a high enough temperature to atomise the molecules, but not so high that the atoms are ionised. Two types of atomiser are commonly used in atomic absorption instruments: flame and electrothermal atomisation.

— Flame-atomisers consist of a nebuliser and a burner. The nebuliser converts the solution into a fine mist or aerosol. The sample is forced through a tiny hole into a chamber through which the oxidant and fuel are flowing. The oxidant and fuel carry the sample into the flame. The burner is usually 5 -10 centimetres long so as to give a long pathlength for the radiation to travel along. The characteristics of the flame can be altered by varying the relative proportions and types of oxidant and fuel used in the flame. Air-acetelyne and Nitrogen oxide-acetylene are the most commonly used mixtures of oxidant and fuel. Thus flames with different temperatures can be produced. This is important because the energy required to cause atomisation, but not ionisation, varies from substance to substance. Instrument manufactures provide guidelines with their instruments about the type of flame to use for specific elements.

— In electrothermal AAS the sample is placed in a small graphite cup which is electrically heated to a temperature (typically 2,000 - 3,000°C) high enough to produce volatilisation and atomisation. The cup is positioned so that the radiation beam passes through the atomised sample. The advantage of electrothermal atomisers is that smaller samples are required and detection limits are lower. Major disadvantages are that they are more expensive to purchase, have a lower sample throughput, are more difficult to operate and have a lower precision than flame-atomisers.

Wavelength selector: A wavelength selector is positioned in the optical path between the flame (or furnace) and the detector. It's purpose is to isolate the spectral line of interest from the rest of the radiation coming from the sample, so that only the radiation of the desired wavelength reaches the detector. Wavelength selectors are typically, monochromatic gratings or filters.

Detector/Read-out: The detector is a photomultiplier tube that converts electromagnetic energy reaching it into an electrical signal. Most modern instruments have a computer to display the signal output and store the spectra.

Atomic Emission Spectroscopy (AES)

Atomic emission spectroscopy (AES) is different from AAS, because it utilises the emission of radiation by a sample, rather than the absorption. For this reason samples usually have to be heated to a higher temperature so that a greater proportion of the atoms are in an excited state (although care must be taken to ensure that ionisation does not occur because the spectra from ionised atoms is different from that of non-ionised atoms). There are a number of ways that the energy can be supplied to a sample, including heat, light, electricity and radio waves.

In AES the sample itself acts as the source of the detected radiation, and therefore there is no need to have a separate radiation source or a chopper. The sample is heated to a temperature where it is atomised and a significant proportion of the atoms is in an excited state.

Atomic emissions are produced when the electrons in an excited state fall back to lower energy levels. Since the allowed energy levels for each atom are different, they each have characteristic emission spectrum from which they can be identified.

Since a food usually contains a wide variety of different minerals, each with a characteristics emission spectrum, the overall spectrum produced contains many absorption peaks. The emitted radiation is therefore passed through a wavelength selector to isolate specific peaks in the spectra corresponding to the atom of interest, and the intensity of the peak is measured using a detector and displayed on a read-out device.

Atomisation-Excitation Source: The purpose of the atomisation-excitation source is to atomise the sample, and to excite the atoms so that they emit a significant amount of detectable radiation. The two most commonly used forms of atomisation-excitation sources in food analysis are Flame and Inductively Coupled Plasma (ICP) devices.

- In flame-AES a nebuliser-burner system is used to atomise the minerals in the sample and excite a large proportion of them to higher energy levels.
- In ICP-AES a special device is used that heats the sample to very high temperatures (6,000 to 10,000 K) in the presence of argon ions. The minerals in the sample are not ionised at these temperatures because of the

high concentration of argon ions (Ar ⇔ Ar^+ + e-) leads to the release of electrons that push the equilibrium towards the non-ionised form of the mineral (M^+ + e- ⇔ M).

Wavelength selectors: Wavelength selectors are used to isolate particular spectral lines, which are characteristic of the material being studied, from all the other spectral lines. A number of different types of wavelength selector are available including filters and gratings.

A filter can only be used to measure the intensity at a particular fixed wavelength, whereas a grating can be used to measure the intensity at many different wavelengths. A filter can therefore only be used to analyse for one type of mineral, whereas a grating can be used to measure many different types of minerals.

References

Diaz, M.; Dunn, C. M.; McClements, D. J.; Decker, E. A., "Use of Caseinophosphopeptides as Natural Antioxidants in Oil-in-Water Emulsions", *J. Agric. Food Chem.*, 2003.

Guzey, D., McClements, D.J. & Weiss, J., "Adsorption kinetics of BSA at air-sugar solution interfaces as affected by sugar type and concentration", *Food Res. Int.*, 2003.

Hu M, McClements DJ, Decker EA, "Lipid oxidation in corn oil-in-water emulsions stabilized by casein, whey protein isolate, and soy protein isolate", *Journal of Agricultural and Food Chemistry*, 51, 1696-1700, 2003.

Kim HJ, Decker EA, McClements DJ., "Influence of sucrose on droplet flocculation in hexadecane oil-in-water emulsions stabilized by beta-lactoglobulin", *J Agr Food Chem.*, 50, 766-772, 2003.

McClements, D.J., "Modulation of Globular Protein Functionality by Weakly Interacting Cosolvents", Crit. Rev. *Food Sci. Nutr.*, 42, 417-471, 2002.

Ogawa, S.; Decker, E. A.; McClements, D. J., "Production and Characterization of O/W Emulsions Containing Cationic Droplets Stabilized by Lecithin-Chitosan Membranes", *J. Agric. Food Chem.*, 2003.

Rampon, V., *et al.*, "Front-Face Fluorescence Spectroscopy Study of Globular Proteins in Emulsions: Displacement of BSA by a Nonionic Surfactant", *J. Agric. Food Chem.*, 2003.

Singh, AP, DJ McClements, AG Marangoni, "Comparison of ultrasonic and pulsed NMR techniques for determination of solid fat content", *J Am Oil Chem Soc*, 79, 431-437, 2002.

Undeland, I., *et al.*, "Consistency and solubility changes in herring (Clupea harengus) light muscle homogenates as a function of pH", *Journal of Agricultural and Food Chemistry*, 51, 3992-3998, 2003.

7

Analysis of Moisture and Total Solids in Food Products

Moisture content is one of the most commonly measured properties of food materials. It is important to food scientists for a number of different reasons:

- *Legal and Labelling Requirements:* There are legal limits to the maximum or minimum amount of water that must be present in certain types of food.
- *Economic:* The cost of many foods depends on the amount of water they contain—water is an inexpensive ingredient, and manufacturers often try to incorporate as much as possible in a food, without exceeding some maximum legal requirement.
- *Microbial Stability:* The propensity of microorganisms to grow in foods depends on their water content. For this reason many foods are dried below some critical moisture content.
- *Food Quality:* The texture, taste, appearance and stability of foods depends on the amount of water they contain.
- *Food Processing Operations:* A knowledge of the moisture content is often necessary to predict the behaviour of foods during processing, e.g. mixing, drying, flow through a pipe or packaging.

It is therefore important for food scientists to be able to reliably measure moisture contents. A number of analytical techniques have been developed for this purpose, which vary in their accuracy, cost, speed, sensitivity, specificity, ease of operation, etc. The choice of an analytical procedure for a particular application depends on the nature of the food being analysed and the reason the information is needed.

The moisture content of a food material is defined through the following equation:

$$\% \, Moisture = (m_w/m_{sample}) \times 100$$

Where m_w is the mass of the water and msample is the mass of the sample. The mass of water is related to the number of water molecules (n_W) by the following expression: $m_w = n_W M_w / N_A$, where M_w is the molecular weight of water (18.0 g per mole) and N_A is Avadagro's number (6.02 10^{23} molecules per mole). In principle, the moisture content of a food can therefore be determined accurately by measuring the number or mass of water molecules present in a known mass of sample.

It is not possible to directly measure the number of water molecules present in a sample because of the huge number of molecules involved. A number of analytical techniques commonly used to determine the moisture content of foods are based on determinations of the mass of water present in a known mass of sample. There are a number of practical problems associated with these techniques that make highly accurate determinations of moisture content difficult or that limit their use for certain applications. For these reasons, a number of other analytical methods have been developed to measure the moisture content of foods that do not rely on direct measurement of the mass of water in a food. Instead, these techniques are based on the fact that the water in a food can be distinguished from the other components in some measurable way.

An appreciation of the principles, advantages and limitations of the various analytical techniques developed to determine the moisture content of foods depends on an understanding of the molecular characteristics of water. A water molecule consists of an oxygen atom covalently bound to two hydrogen atoms (H_2O). Each of the hydrogen atoms has a small positive charge (δ+), while the oxygen atom has two lone pairs of electrons that each has a small negative charge (δ-). Consequently, water molecules are capable of forming relatively strong hydrogen bonds ($O\text{-}H^{\delta+} \leftrightarrow {}^{\delta-}O$) with four neighbouring water molecules.

The strength and directionality of these hydrogen bonds are the origin of many of the unique physicochemical properties of water. The development of analytical techniques to determine the moisture content of foods depends on being able to distinguish water (the "analyte") from the other components in the food (the "matrix"). The characteristics of water that are most commonly used to achieve this are: its relatively low boiling point; its high polarity; its ability to undergo unique chemical reactions with certain reagents; its unique electromagnetic absorption spectra; and, its characteristic physical properties (density, compressibility, electrical conductivity and refractive index).

Despite having the same chemical formula (H_2O) the water molecules in a food may be present in a variety of different molecular environments depending on their interaction with the surrounding molecules. The water molecules in these different environments normally have different physiochemical properties:

- *Bulk water:* Bulk water is free from any other constituents, so that each water molecule is surrounded only by other water molecules. It therefore has physicochemical properties that are the same as those of pure water, e.g., melting point, boiling point, density, compressibility, heat of vaporisation, electromagnetic absorption spectra.
- *Capillary or trapped water:* Capillary water is held in narrow channels between certain food components because of capillary forces. Trapped water is held within spaces within a food that are surrounded by a physical barrier that prevents the water molecules from easily escaping, e.g., an emulsion droplet or a biological

cell. The majority of this type of water is involved in normal water-water bonding and so it has physicochemical properties similar to that of bulk water.

— *Physically bound water:* A significant fraction of the water molecules in many foods are not completely surrounded by other water molecules, but are in molecular contact with other food constituents, e.g. proteins, carbohydrates or minerals. The bonds between water molecules and these constituents are often significantly different from normal water-water bonds and so this type of water has different physicochemical properties than bulk water e.g., melting point, boiling point, density, compressibility, heat of vaporisation, electromagnetic absorption spectra.

— *Chemically bound water:* Some of the water molecules present in a food may be chemically bonded to other molecules as water of crystallisation or as hydrates, e.g. $NaSO_4.10H_20$. These bonds are much stronger than the normal water-water bond and therefore chemically bound water has very different physicochemical properties to bulk water, e.g., lower melting point, higher boiling point, higher density, lower compressibility, higher heat of vaporisation, different electromagnetic absorption spectra.

Foods are heterogeneous materials that contain different proportions of chemically bound, physically bound, capillary, trapped or bulk water. In addition, foods may contain water that is present in different physical states: gas, liquid or solid. The fact that water molecules can exist in a number of different molecular environments, with different physicochemical properties, can be problematic for the food analyst trying to accurately determine the moisture content of foods.

Many analytical procedures developed to measure moisture content are more sensitive to water in certain types of molecular environment than to water in other types of molecular environment. This means that the measured value of the moisture content of a particular food may depend on the experimental technique used to carry out the measurement. Sometimes food analysts are interested in determining the amounts of water in specific molecular environments, rather than the total water content. For example, the rate of microbial growth in a food depends on the amount of bulk water present in a food, and not necessarily on the total amount of water present. There are analytical techniques available that can provide some information about the relative fractions of water in different molecular environments.

Sample Preparation

Selection of a representative sample, and prevention of changes in the properties of the sample prior to analysis, are two major potential sources of error in any food analysis procedure. When determining the moisture content of a food it is important to prevent any loss or gain of water. For this reason, exposure of a sample to the atmosphere, and excessive temperature fluctuations, should be minimised. When samples are stored in containers it is common practice to fill the container to the top to prevent a large headspace, because this reduces changes in the sample due to equilibration with its environment.

Use of Evaporation Methods

Principles

These methods rely on measuring the mass of water in a known mass of sample. The moisture content is determined by measuring the mass of a food before and after the water is removed by evaporation:

$$\%\,Moisture = \frac{M_{INITIAL} - M_{DRIED}}{M_{INITIAL}} \times 100$$

Here, $M_{INITIAL}$ and M_{DRIED} are the mass of the sample before and after drying, respectively. The basic principle of this technique is that water has a lower boiling point than the other major components within foods, e.g., lipids, proteins, carbohydrates and minerals. Sometimes a related parameter, known as the total solids, is reported as a measure of the moisture content. The total solids content is a measure of the amount of material remaining after all the water has been evaporated:

$$\% \text{ Total Solids} = \frac{M_{DRIED}}{M_{INITIAL}} \times 100$$

Thus, % Total solids = (100 - % Moisture). To obtain an accurate measurement of the moisture content or total solids of a food using evaporation methods it is necessary to remove all of the water molecules that were originally present in the food, without changing the mass of the food matrix. This is often extremely difficult to achieve in practice because the high temperatures or long times required to remove all of the water molecules would lead to changes in the mass of the food matrix. For this reason, the drying conditions used in evaporation methods are usually standardised in terms of temperature and time so as to obtain results that are as accurate and reproducible as possible given the practical constraints. Using a standard method of sample preparation and analysis helps to minimise sample-to-sample variations within and between laboratories.

Evaporation Devices

The thermal energy used to evaporate the water from a food sample can be provided directly or indirectly.

Convection and forced draft ovens

Weighed samples are placed in an oven for a specified time and temperature and their dried mass is determined, or they are dried until they reach constant mass. The thermal energy used to evaporate the water is applied directly to the sample via the shelf and air that surround it. There are often considerable temperature variations within convection ovens, and so precise measurements are carried out using forced draft ovens that circulate the air so as to achieve a more uniform temperature distribution within the oven. Samples that contain significant quantities of carbohydrates that might undergo chemical changes or volatile materials other than water should not be dried in a convection or forced draft oven. Many official methods of analysis are based on forced draft ovens.

Vacuum oven

Weighed samples are placed under reduced pressure (typically 25-100 mm Hg) in a vacuum oven for a specified time and temperature and their dried mass is determined. The thermal energy used to evaporate the water is applied directly to the sample via the metallic shelf that it sits upon. There is an air inlet and outlet to carry the moisture lost from the sample out of the vacuum oven, which prevents the accumulation of moisture within the oven. The boiling point of water is reduced when it is placed under vacuum. Drying foods in a vacuum oven therefore has a number of advantages over conventional oven drying techniques. If the sample is heated at the same temperature, drying can be carried out much quicker. Alternatively, lower temperatures can be used to remove the moisture (e.g. 70°C instead of 100°C), and so problems associated with degradation of heat labile substances can be reduced. A number of vacuum oven methods are officially recognised.

Microwave oven

Weighed samples are placed in a microwave oven for a specified time and power-level and their dried mass is weighed. Alternatively, weighed samples may be dried until they reach a constant final mass - analytical microwave ovens containing balances to continuously monitor the weight of a food during drying are commercially available. The water molecules in the food evaporate because they absorb microwave energy, which causes them to become thermally excited. The major advantage of microwave methods over other drying methods is that they are simple to use and rapid to carry out. Nevertheless, care must be taken to standardise the drying procedure and ensure that the microwave energy is applied evenly across the sample. A number of microwave oven drying methods are officially recognised.

Infrared lamp drying

The sample to be analysed is placed under an infrared lamp and its mass is recorded as a function of time. The water molecules in the food evaporate because they absorb infrared energy, which causes them to become thermally excited. One of the major advantages of infrared drying methods is that moisture contents can be determined rapidly using inexpensive equipment,. This is because the IR energy penetrates into the sample, rather than having to be conducted and convected inwards from the surface of the sample. To obtain reproducible measurements it is important to control the distance between the sample and the IR lamp and the dimensions of the sample. IR drying methods are not officially recognised for moisture content determinations because it is difficult to standardise the procedure. Even so, it is widely used in industry because of its speed and ease of use.

Practical Considerations

— *Sample dimensions:* The rate and extent of moisture removal depends on the size and shape of the sample, and how finely it is ground. The greater the surface area of material exposed to the environment, the faster the rate of moisture removal.

— *Clumping and surface crust formation:* Some samples tend to clump together or form a semipermeable surface crust during the drying procedure. This can lead to erroneous and irreproducible results because the loss of moisture is restricted by the clumps or crust. For this reason samples are often mixed with dried sand to prevent clumping and surface crust formation.

— *Elevation of boiling point:* Under normal laboratory conditions pure water boils at 100 °C. Nevertheless, if solutes are present in a sample the boiling point of water is elevated. This is because the partial vapour pressure of water is decreased and therefore a higher temperature has to be reached before the vapour pressure of the system equals the atmospheric pressure. Consequently, the rate of moisture loss from the sample is slower than expected. The boiling point of water containing solutes (T_b) is given by the expression, $T_b = T_0 + 0.51_m$, where T_0 is the boiling point of pure water and m is the molality of solute in solution (mol/kg of solvent).

— *Water type:* The ease at which water is removed from a food by evaporation depends on its interaction with the other components present. Free water is most easily removed from foods by evaporation, whereas more severe conditions are needed to remove chemically or physically bound water. Nevertheless, these more extreme conditions can cause problems due to degradation of other ingredients which interfere with the analysis.

— *Decomposition of other food components:* If the temperature of drying is too high, or the drying is carried out for too long, there may be decomposition of some of the heat-sensitive components in the food. This will cause a change in the mass of the food matrix and lead to errors in the moisture content determination. It is therefore normally necessary to use a compromise time and temperature, which are sufficient to remove most of the moisture, but not too long to cause significant thermal decomposition of the food matrix. One example of decomposition that interferes with moisture content determinations is that of carbohydrates.

$$C_6H_{12}O_6 \xrightarrow{\text{heat}} 6C + 6\,H_2O$$

The water that is released by this reaction is not the water we are trying to measure and would lead to an overestimation of the true moisture content. On the other hand, a number of chemical reactions that occur at elevated temperatures lead to water absorption, e.g., sucrose hydrolysis (sucrose + $H_2O \xrightarrow{heat}$ fructose + glucose), and therefore lead to an underestimation of the true moisture content. Foods that are particularly susceptible to thermal decomposition should be analysed using alternative methods, e.g. chemical or physical.

— *Volatilisation of other food components:* It is often assumed that the weight loss of a food upon heating is entirely due to evaporation of the water. In practice, foods often contain other volatile constituents that can also be lost during heating, e.g., flavours or odours. For most foods, these volatiles only make up a very small proportion and can therefore be ignored. For foods that do contain significant amounts of volatile components (e.g. spices and herbs) it is necessary to use alternative methods to determine their moisture content, e.g., distillation, chemical or physical methods.

— *High moisture samples:* Food samples that have high moisture contents are usually dried in two stages to prevent "spattering" of the sample, and accumulation of moisture in the oven. Spattering is the process whereby some of the water jumps out of the food sample during drying, carrying other food constituents with it. For example, most of the moisture in milk is removed by heating on a steam bath prior to completing the drying in an oven.

— *Temperature and power level variations:* Most evaporation methods stipulate a definite temperature or power level to dry the sample so as to standardise the procedure and obtain reproducible results. In practice, there are often significant variations in temperatures or power levels within an evaporation instrument, and so the efficiency of the drying procedure depends on the precise location of the sample within the instrument. It is therefore important to carefully design and operate analytical instruments so as to minimise these temperature or power level variations.

— *Sample pans:* It is important to use appropriate pans to contain samples, and to handle them correctly, when carrying out a moisture content analysis. Typically aluminium pans are used because they are relatively cheap and have a high thermal conductivity. These pans usually have lids to prevent spattering of the sample, which would lead to weight loss and therefore erroneous results. Pans should be handled with tongs because fingerprints can contribute to the mass of a sample. Pans should be dried in an oven and stored in a desiccator prior to use to ensure that no residual moisture is attached to them.

Use of Distillation Methods

Distillation methods are based on direct measurement of the amount of water removed from a food sample by evaporation: %Moisture = 100 $(M_{WATER}/M_{INITIAL})$. In contrast, evaporation methods are based on indirect measurement of the amount of water removed from a food sample by evaporation: %Moisture = 100 $(M_{INITIAL} - M_{DRIED})/M_{INITIAL}$. Basically, distillation methods involve heating a weighed food sample $(M_{INITIAL})$ in the presence of an organic solvent that is immiscible with water. The water in the sample evaporates and is collected in a graduated glass tube where its mass is determined (M_{WATER}).

Dean and Stark Method

Distillation methods are best illustrated by examining a specific example: the Dean and Stark method. A known weight of food is placed in a flask with an organic solvent such as xylene or toluene. The organic solvent must be insoluble with water; have a higher boiling point than water; be less dense than water; and be safe to use. The flask containing the sample and the organic solvent is attached to a condenser by a side arm and the mixture is heated. The water in the sample evaporates and moves up into the condenser where it is cooled and converted back into liquid water, which then trickles into the graduated tube. When no more water is collected in the graduated tube, distillation is stopped and the volume of water is read from the tube.

Practical Considerations

There are a number of practical factors that can lead to erroneous results:

- emulsions can sometimes form between the water and the solvent which are difficult to separate;
- water droplets can adhere to the inside of the glassware,
- decomposition of thermally labile samples can occur at the elevated temperatures used.

Advantages and Disadvantages

- Advantages: Suitable for application to foods with low moisture contents; Suitable for application to foods containing volatile oils, such as herbs or spices, since the oils remain dissolved in the organic solvent, and therefore do not interfere with the measurement of the water; Equipment is relatively cheap, easy to setup and operate; Distillation methods have been officially sanctioned for a number of food applications.
- Disadvantages: Destructive; Relatively time-consuming; Involves the use of flammable solvents; Not applicable to some types of foods.

Chemical Reaction Methods

Reactions between water and certain chemical reagents can be used as a basis for determining the concentration of moisture in foods. In these methods a chemical reagent is added to the food that reacts specifically with water to produce a measurable change in the properties of the system, e.g., mass, volume, pressure, pH, colour, conductivity. Measurable changes in the system are correlated to the moisture content using calibration curves.

To make accurate measurements it is important that the chemical reagent reacts with all of the water molecules present, but not with any of the other components in the food matrix. Two methods that are commonly used in the food industry are the Karl-Fisher titration and gas production methods. Chemical reaction methods do not usually involve the application of heat and so they are suitable for foods that contain thermally labile substances that would change the mass of the food matrix on heating or foods that contain volatile components that might be lost by heating.

Karl-Fisher method

The Karl-Fisher titration is often used for determining the moisture content of foods that have low water contents (e.g. dried fruits and vegetables, confectionary, coffee, oils and fats). It is based on the following reaction:

$$2H_2O + SO_2 + I_2 \rightarrow H_2SO_4 + 2HI$$

This reaction was originally used because HI is colourless, whereas I_2 is a dark reddish brown colour, hence there is a measurable change in colour when water reacts with the added chemical reagents. Sulphur dioxide and iodine are gaseous and would normally be lost from solution. For this reason, the above reaction has been modified by adding solvents (e.g., C5H5N) that keep the S_2O and I_2 in solution, although the basic principles of the method are the same. The food to be analysed is placed in a beaker containing solvent and is then titrated with Karl Fisher reagent (a solution that contains iodine).

While any water remains in the sample the iodine reacts with it and the solution remains colourless (HI), but once all the water has been used up any additional iodine is observed as a dark red brown colour. The volume of iodine solution required to titrate the water is measured and can be related to the moisture content using a pre-prepared calibration curve. The precision of the technique can be improved by using electrical methods to follow the end-point of the reaction, rather than observing a colour change. Relatively inexpensive commercial instruments have been developed which are based on the Karl-Fisher titration, and some of these are fully automated to make them less labour intensive.

Gas Production Methods

Commercial instruments are also available that utilise specific reactions between chemical reagents and water that lead to the production of a gas. For example, when a food sample is mixed with powdered calcium carbide the amount of acetylene gas produced is related to the moisture content.

$$CaC_2 + 2H_2O \rightarrow C_2H_2(gas) + Ca(OH)_2$$

The amount of gas produced can be measured in a number of different ways, including (i) the volume of gas produced, (ii) the decrease in the mass of the sample after the gas is released, and (iii) the increase in pressure of a closed vessel containing the reactants.

Physical Methods

A number of analytical methods have been developed to determine the moisture content of foods that are based on the fact that water has appreciably different bulk physical characteristics than the food matrix, e.g. density, electrical conductivity or refractive index. These methods are usually only suitable for analysis of foods in which the composition of the food matrix does not change significantly, but the ratio of water-to-food matrix changes. For example, the water content of oil-in-water emulsions can be determined by measuring their density or electrical conductivity because the density and electrical conductivity of water are significantly higher than those of oil.

If the composition of the food matrix changes as well as the water content, then it may not be possible to accurately determine the moisture content of the food because more than one food composition may give the same value for the physical property being measured. In these cases, it may be possible to use a combination of two or more physical methods to determine the composition of the food, e.g., density measurements in combination with electrical conductivity measurements.

Spectroscopic Methods

Spectroscopic methods utilise the interaction of electromagnetic radiation with materials to obtain information about their composition, e.g., X-rays, UV-visible, NMR, microwaves and IR. The spectroscopic methods developed to measure the moisture content of foods are based on the fact that water absorbs electromagnetic radiation at characteristic wavelengths that are different from the other components in the food matrix. The most widely used physical methods are based on measurements of the absorption of microwave or infrared energy by foods.

Microwave and infrared radiation are absorbed by materials due to their ability to promote the vibration and/or rotation of molecules. The analysis is carried out at a wavelength where the water molecules absorb radiation, but none of the other components in the food matrix do. A measurement of the absorption of radiation at this wavelength can then be used to determine the moisture content: the higher the moisture content, the greater the absorption. Instruments based on this principle are commercially available and can be used to determine the moisture content in a few minutes or less.

It is important not to confuse infrared and microwave absorption methods with infrared lamp and microwave evaporation methods. The former use low energy waves that cause no physical or chemical changes in the food, whereas the latter use high-energy waves to evaporate the water. The major advantage of these methods is that they are capable of rapidly determining the moisture content of a food with little or no sample preparation and are therefore particularly useful for quality control purposes or rapid measurements of many samples.

Determine Water in Different Molecular Environments

The overall water content of a food is sometimes not a very reliable indication of the quality of a food because the water molecules may exist in different environments within foods, e.g., "bound" or "free". Here "bound water" refers to water that is physically or chemically bound to other food components, whereas "free water" refers to bulk, capillary or entrapped water. For example, the microbial stability or physicochemical properties of a food are often determined by the amount of free water present, rather than by the total amount of water present. For this reason, it is often useful for food scientists to be able to determine the amount of water in different molecular environments within a food. A variety of analytical methods are available that can provide this type of information.

Vapour Pressure Methods

A physical parameter that is closely related to the amount of free water present in a food is the water activity.

$$a_w = \frac{P}{P_0}$$

where, P is the partial pressure of the water above the food and P_0 is the vapour pressure of pure water at the same temperature. Bound water is much less volatile than free water, and therefore the water activity gives a good indication of the amount of free water present. A variety of methods are available for measuring the water activity of a sample based on its vapour pressure. Usually, the sample to be analysed is placed in a closed container and allowed to come into equilibrium with its environment. The water content in the headspace above the sample is then measured and compared to that of pure water under the same conditions.

Thermogravimetric Methods

Thermogravimetric techniques can be used to continuously measure the mass of a sample as it is heated at a controlled rate. The temperature at which water evaporates depends on its molecular environment: free water normally evaporates at a lower temperature than bound water. Thus by measuring the change in the mass of a sample as it loses water during heating it is often possible to obtain an indication of the amounts of water present in different molecular environments.

Calorimetric Methods

Calorimetric techniques such as differential scanning calorimetry (DSC) and differential thermal analysis (DTA) can be used to measure changes in the heat absorbed or released by a material as its temperature is varied at a controlled rate. The melting point of water depends on its molecular environment: free water normally melts at a higher temperature than bound water. Thus by measuring the enthalpy change of a sample with temperature it is possible to obtain an indication of the amounts of water present in different molecular environments.

References

Diaz, M.; Dunn, C. M.; McClements, D. J.; Decker, E. A., "Use of Caseinophosphopeptides as Natural Antioxidants in Oil-in-Water Emulsions", *J. Agric. Food Chem.*, 2003.

Guzey, D., McClements, D.J. & Weiss, J., "Adsorption kinetics of BSA at air-sugar solution interfaces as affected by sugar type and concentration", *Food Res. Int.*, 2003.

Hu M, McClements DJ, Decker EA, "Lipid oxidation in corn oil-in-water emulsions stabilized by casein, whey protein isolate, and soy protein isolate", *Journal of Agricultural and Food Chemistry*, 51, 1696-1700, 2003.

Kim HJ, Decker EA, McClements DJ., "Influence of sucrose on droplet flocculation in hexadecane oil-in-water emulsions stabilized by beta-lactoglobulin", *J Agr Food Chem.*, 50, 766-772, 2003.

Ogawa, S.; Decker, E. A.; McClements, D. J., "Production and Characterization of O/W Emulsions Containing Cationic Droplets Stabilized by Lecithin-Chitosan Membranes", *J. Agric. Food Chem.*, 2003.

Rampon, V., *et al.*, "Front-Face Fluorescence Spectroscopy Study of

Globular Proteins in Emulsions: Displacement of BSA by a Nonionic Surfactant", *J. Agric. Food Chem.*, 2003.

Undeland, I., *et al.*, "Consistency and solubility changes in herring (Clupea harengus) light muscle homogenates as a function of pH", *Journal of Agricultural and Food Chemistry*, 51, 3992-3998, 2003.

8

Instrumental Techniques for Food Analysis

Temperature changes cause alterations in the physical and chemical properties of food components which influence the overall properties of the final product, e.g., taste, appearance, texture and stability. Chemical reactions such as hydrolysis, oxidation or reduction may be promoted, or physical changes, such as evaporation, melting, crystallization, aggregation or gelation may occur.

A better understanding of the influence of temperature on the properties of foods enables food manufacturers to optimise processing conditions and improve product quality. It is therefore important for food scientists to have analytical techniques to monitor the changes that occur in foods when their temperature varies. These techniques are often grouped under the general heading of thermal analysis.

In principle, most analytical techniques can be used, or easily adapted, to monitor the temperature-dependent properties of foods, e.g., spectroscopic (NMR, UV-visible, IR spectroscopy, fluorescence), scattering (light, X-rays, neutrons), physical (mass, density, rheology, heat capacity) etc. Nevertheless, at present the term thermal analysis is usually reserved for a narrow range of techniques that measure changes in the physical properties of foods with temperature, e.g., mass, density, rheology, heat capacity. For this reason, only these techniques will be considered in this lecture.

Temperature Dependent Properties

Initially, it is useful to highlight some of the physical changes that occur in food components when the temperature is varied.

Density of Pure Materials

The density of pure materials, which do not undergo phase transitions, usually decrease as the temperature is increased. This is because the atoms in the material move around more vigorously when they gain thermal energy, and so the space between the molecules increases. The mass of a material is independent of temperature, and so an increase in volume with temperature leads to a decrease in density (since = m/V). Knowledge of the temperature-dependence of the density of a food material is often used by engineers to design processing operations, e.g., containers for storing materials or pipes through which materials flow.

In materials that do undergo phase transitions the variation of the density with temperature is more dramatic. A solid usually has a higher density than a liquid, and so when a solid melts or a liquid crystallises there is a significant change in density superimposed on the normal variation of density with temperature. The use of density measurements to monitor melting and crystallisation of materials will be discussed later.

Phase Transitions

The term phase transition refers to the process whereby a material is converted from one physical state to another. The most commonly occurring phase transitions in foods are melting (solid-to-liquid), crystallisation (liquid-to-solid), evaporation (liquid-to-gas), condensation (gas-to-liquid), sublimation (solid-to-gas) and glass transitions (glassy-to-rubbery). When a material changes from one physical state to another it either absorbs or gives out heat. A process that absorbs heat is an endothermic process, whereas a process that evolves heat is an exothermic process.

The overall properties of foods may be drastically altered when key components undergo phase transitions, and so it is important to have analytical techniques for monitoring these processes. These techniques utilise measurements of physical properties of a material that change when a material undergoes a phase transition, e.g., molecular structure, molecular mobility, density, rheology, heat capacity.

Gelation

Many foods contain components that are capable of forming a gel when the food is heated or cooled under appropriate conditions. Most food gels are three-dimensional networks of aggregated or entangled biopolymers or colloidal particles that entrap a large volume of water, to give the whole structure "solid-like" characteristics. The physical properties of gels, such as appearance (transparent or opaque), water holding capacity, rheology and stability, depend ultimately on the type, structure and interactions of the molecules or particles that they contain.

Common examples of foods in which gelation makes an important contribution to their overall properties are eggs, starches, jellies, yogurts and meat products. In some foods a gel is formed on heating (heat-setting gels), whilst in others it is formed on cooling (cold-setting gels). Gels may also be either thermo-reversible or thermo-irreverisble, depending on whether gelation is reversible or not. Gelatin is an example of a cold-setting thermo-reversible gel: when a solution of gelatin molecules is cooled below a certain temperature a gel is formed, but when it is reheated the gel melts. Egg-white is an example of a heat-setting thermo-irreverisble gel.

When an egg is heated above a temperature where gelation occurs a characteristic white gel is formed, however, when the egg is cooled back to room temperature the gel remains white, i.e., it doesn't revert back into the liquid from which it was formed. For ingredients that gel it is important to know the temperature at which gelation occurs, the gelation rate, and the nature of the gel formed. Thus thermal analytical techniques are needed by food scientist to measure these properties.

Analytical Techniques

A variety of different analytical techniques have been developed to monitor changes in the physical properties of food components that occur in response to controlled changes in temperature. A number of the most important of these thermal analysis techniques are described below.

Thermogravimetric Techniques

Thermogravimetric techniques continuously measure the mass of a sample as it is heated or cooled at a controlled rate, or is held at a particular temperature for a period of time. Thermogravimetry is useful for monitoring processes that involve a change in the mass of a food or food component, e.g., drying, liberation of gasses, absorption of moisture. To mimic the various types of processing and storage conditions that a food might normally experience, thermogravimetric instruments have been specially designed to allow measurements to be carried out under specific environments, e.g., controlled pressures or atmospheres.

Gravimetric instruments typically consist of a sensitive balance situated within a container whose pressure, temperature and gaseous environment can be carefully controlled. The mass of a sample may either increase or decrease with temperature or time depending on the specific physicochemical processes occurring. Heating often leads to a reduction in mass because of evaporation of volatile components and various chemical reactions that liberate gasses.

On the other hand, the mass of a food may increase due to absorption of moisture from the atmosphere. The ability to be able to carefully control the temperature, pressure and composition of the gasses surrounding a sample is extremely valuable for food scientists, because it allows them to model processes such as drying, cooking, and uptake of moisture during storage.

Dilatometry Measurements

A dilatometer is a device that is used to measure the change in density of a material as a function of time or temperature. Dilatometry measurements are routinely used for monitoring the crystallisation and melting of fats in foods. A weighed amount of melted fat is poured into a graduated glass U-tube that is thermostatted in a temperature controlled water bath. The sample is then cooled at a controlled rate and the change in volume of the material is measured as a function of temperature.

The density of a solid is usually greater than that of a liquid, thus the volume of a sample decreases when crystallisation occurs, and increases when melting occurs. Dilatometry can therefore be used to provide information about the melting and crystallisation of fatty foods. For food scientists, the most important information is the temperature at which melting or crystallisation begins, the temperature range over which the phase transition occurs, and the value of the solid fat content at any particular temperature.

Rheological Techniques

Rheology is the study of the deformation and flow of matter. Rheological techniques used for thermal analysis measure the change in the rheological characteristics of a sample as a function of temperature. A sample is usually contained in a measurement cell whose temperature can be varied in a systematic fashion. A stress is applied to the sample and the resulting strain is measured. The relationship between the stress and strain gives information about the rheological properties of the material being tested.

The stress can be applied to a material in a number of different ways, depending on the type of information required. The stresses used are normally small enough to prevent any changes in the properties of the material during the test. If large stresses were applied to a material they might promote structure breakdown, which would alter the rheological properties of the material during the test.

Rheological thermal analysis techniques are often used to monitor the temperature dependent rheological properties of liquids, gels and solids. For example, they are commonly used to monitor the temperature dependence of the shear modulus of fatty foods, the viscosity of biopolymer solutions, and the shear modulus of biopolymer gels. These techniques provide useful information about the temperature at which thermal transitions occur, the rate at which these changes occur and the final rheological properties of the food. This type of information is used by food scientists to design foods with improved properties, and to optimise processing conditions.

DTA and DSC Techniques

Differential Thermal Analysis (DTA) and Differential Scanning Calorimetry (DSC) techniques rely on changes in the heat absorbed or released by a material as its temperature is varied at a controlled rate. These changes occur when components within a food undergo some type of phase transition or chemical reaction.

Differential Thermal Analysis Technique

DTA is defined as "a technique for recording the difference in temperature between a substance and a reference material against time or temperature as the two specimens are subjected to identical temperature regimes in an environment heated or cooled at a controlled rate". A typical instrument consists of two measurement cells that are located in a temperature-controlled environment, whose temperature can be varied in a controlled fashion.

The sample to be tested is placed into the "sample cell", while a reference material of known thermal properties (often distilled water) is placed in the "reference cell". The two cells are then heated or cooled together at a controlled rate. The small difference in temperature between the "sample cell" and "reference cell" $\Delta T = T_{sample} - T_{reference}$ is measured using accurate thermocouples placed below the cells as the temperature of the external environment ($T_{external}$) is varied in a controlled fashion.

The output of the instrument is therefore a plot of T versus Texternal. Information about thermal transitions that occur within a sample can be obtained by analysing the T versus Texternal thermogram. If the temperature of the "sample cell" is greater than that of the "reference cell" ($\Delta T > 0$), then the sample has undergone an exothermic reaction, i.e., it has given out heat.

Conversely, if the temperature of the "reference cell" is greater than that of the "sample cell" ($\Delta T < 0$), then the sample has undergone an endothermic reaction, i.e., it has adsorbed heat. The nature of a peak (exothermic, endothermic, shape) provides information about the type of transition(s) occurring. The position of the peak provides information about the temperature that the transition occurs. The area under a peak depends on the amount of material involved in the transition and the enthalpy change per unit amount of material.

Differential Scanning Calorimetry Technique

DSC is a technique for recording the energy required to keep a zero temperature difference between a sample cell and a reference cell which are either heated or cooled at a controlled rate. The thermocouples constantly measure the temperature of each cell and heaters supply heat to one or other of the cells so that they both have exactly the same temperature. If a sample were to undergo a phase transition it would either absorb or release heat.

To keep the temperature of the two samples the same an equivalent amount of energy must be supplied to either the test or reference cells. Special electrical circuitry is used to determine the amount of energy needed to keep the two measurement cells at the same temperature. DSC data is therefore reported as the rate of energy absorption (Q) by the sample relative to the reference material as a function of the external temperature. Information about thermal transitions that occur within a sample are obtained by analysing the Q versus $T_{external}$ thermogram. It should be noted that it is also possible to measure the change in the heat released by a material as a function of time under isothermal (constant temperature) conditions.

Isothermal Titration Calorimetry (ITC)

ITC is used to measure enthalpy changes that occur as the result of interactions between different types of molecules. An ITC instrument consists of a reference cell, a sample cell and an injector. A reference material, that does not undergo any enthalpy changes during the experiment is placed in the reference cell. A solution of one type of molecule is placed in the sample cell ("sample solution"), and a solution of another type of molecule is placed in the injector ("injection solution").

Small aliquots of the injection solution are then injected periodically into the sample solution contained within the sample cell, and the energy required to keep the sample and reference cells at the same temperature is measured as a function of time. The resulting thermogram consists of a plot of Q versus time, which consists of a series of enthalpy peaks corresponding to the series of injections. By analysing the nature (exothermic, endothermic), magnitude (area under the curve) and shape of the peaks it is possible to obtain valuable information about interactions between molecules in the injector and in the sample cell.

Applications

Specific heat capacity

The specific heat capacity is an important quantity in the food industry because it determines the amount of energy that must be supplied or withdrawn from a material in order to increase or decrease its temperature by a given amount. Knowledge of the specific heat capacity of a material is therefore important in the design of processes such as chilling, freezing, warming, sterilisation and cooking.

DSC and DTA can be used to measure the specific heat capacities of food materials. A known mass of material is placed in a sample cell, which is then heated or cooled at a controlled rate. For DSC, the specific heat capacity is determined from the equation: Q = m CP dT/dt, where Q is the heat flow per unit time, m is the sample mass, CP is the specific heat capacity of the material, and dT/dt is the rate of change of the external temperature.

Phase transitions

DSC and DTA are routinely used in the food industry to characterise phase transitions in foods, e.g. crystallisation, melting, glass transitions and conformational changes. They can be used to provide information about the temperature at which transitions occur (Ttr), the enthalpy change associated with a transition (Htr), the type of transition involved (exothermic or endothermic), and the quantify of material that undergoes a transition.

When a material changes its physical state from solid-to-liquid (melting) or from liquid-to-solid (crystallisation) it absorbs or gives out heat, respectively. A process that absorbs heat is an endothermic process, whereas a process that evolves heat is an exothermic process. Pure substances usually have very sharp melting or crystallisation points and therefore all the heat is absorbed or evolved over a narrow range of temperatures, leading to a sharp DSC or DTA peak.

Many food components are chemically complex materials and therefore the phase transitions occur over a wide range of temperatures, e.g. edible oils contain a wide variety of different triacylglycerols each with its own melting point. Peaks from food oils may also be complicated by the fact that triacylglycerols can crystallise in more than one different crystalline structure, i.e., they are polymorphic.

Molecular interactions

ITC can be used to provide valuable information about interactions between different types of molecules, e.g., binding interactions or conformational changes. As an example, we will consider the use of ITC for quantifying the binding of a ligand molecule (L) to a protein molecule (P): P + L PL. A solution containing the ligand is placed into the injector, while a solution containing the protein is placed into the sample cell. Small aliquots of the ligand solution are then injected into the sample solution at regular intervals.

The interval between each injection should be long enough to allow any reactions to go to completion. The instrument records the enthalpy change that occurs after each injection as a result of the interaction between the ligand and protein molecules. By measuring the change in the enthalpy with ligand concentration in the sample cell it is possible to obtain information about the number of binding sites on the protein, the strength of the binding interaction and the thermodynamics of the binding interaction.

Spectroscopic Techniques

A variety of the instruments that are commonly used to analyse food materials are based on spectroscopy, e.g., UV-visible, fluorescence, atomic, infrared and nuclear magnetic resonance spectroscopies. These instruments utilise interactions between electromagnetic radiation and matter to provide information about food properties, e.g., molecular composition, structure, dynamics and interactions. An appreciation of the operating principles of these instruments depends on an understanding of the distribution of energy within atoms and molecules, of the characteristics of electromagnetic radiation, and of the interaction of electromagnetic radiation with atoms and molecules.

Energy Distribution in Atoms and Molecules

Atoms and molecules can only exist in a limited number of discrete energy levels: they cannot have energies between these levels, i.e., their energy levels are quantised. Each molecular species has a unique set of energy levels that depends on its unique atomic structure (electrons, protons, neutrons) and molecular structure (type and arrangement of atoms and bonds).

The lowest of these energy levels is referred to as the ground state, while higher levels are referred to as excited states. The potential energy of an atom or molecule is usually defined relative to the ground state (which is arbitrarily taken to have zero energy). The potential energy of a molecule is made-up of contributions from a number of different sources: electronic, vibrational, rotational, translation and nuclear.

- *Electronic Energy Levels*: Electrons in an atom are arranged into a number of different shells and sub-shells. An electron can move from one of these sub-shell levels to another by absorbing or emitting radiation of an appropriate energy. The system is then said to have undergone an electronic transition. Electronic transitions may involve electrons that are in inner shells (higher energy) or outer shells (lower energy) of atoms.
- *Vibrational Energy Levels*: Molecules (but not atoms) can vibrate in a number of different modes, e.g., the atoms can compress or stretch along the axis of a bond, or they can bend symmetrically or asymmetrically. Each of these vibrations occurs at a characteristic frequency (energy) which depends on the mass of the atoms and the strength of the bonds involved.
- *Rotational Energy Levels*: Molecules often contain chemical groups that are capable of rotating around certain bonds at fixed frequencies (and therefore energies). Each group has a specific number of frequencies at which it rotates and therefore has a specific number of quantised rotational energy levels. The rotation frequency is determined by the mass of the atoms involved and their distance from the axis of rotation.
- *Nuclear Energy Levels*: The nuclei of certain atoms have a property known as spin. A (charged) spinning nucleus generates a small magnetic field and can be thought of as being a small magnet. Normally, this magnet can be orientated in any direction, but in the presence of an external magnetic field it can only align itself either with or against the field, i.e., it is quantised. Transitions between the different energy levels within the nuclei

can be made to occur by applying radiation of a specific energy to the sample.

— *Translational Energy Levels*: Atoms and molecules are in continual translational motion because of the thermal energy of the system. Translational energy levels are quantised, however, the differences between the energy levels are so small that the molecules act as though the energy is distributed continuously.

Electromagnetic Waves

Electromagnetic waves may be thought of as particles of energy (photons) that move through space with wave-like properties, i.e., they exhibit wave-particle duality. They consist of oscillating electric and magnetic fields that are perpendicular to one another, and to the direction of propagation.

The sinusoidal variation in the amplitude of the electric vector of the wave can be plotted as a function of time (at a fixed position within a material) or as a function of distance (at a fixed point in time). A monochromatic (single wavelength) electromagnetic wave that propagates through a vacuum can be described completely by its frequency, wavelength and amplitude (or parameters derived from these):

— The frequency (ν) of a wave is the number of cycles per second (Hz = s^{-1}).

— The period (T) of a wave is the time taken to complete a cycle: $T = 1/\nu$.

— The wavelength (λ) is the distance between successive maxima of a wave.

— The wave number is the number of cycles per unit distance ($=1/\lambda$).

— The amplitude (A) of a wave is the maximum magnitude of the electric vector.

— The intensity (I) of a wave is proportional to the square of the amplitude. It is the amount of energy passing through a given area per second. Increasing the intensity of an electromagnetic wave increases the number of quanta passing a given area per second, not the energy of each individual quantum.

— The velocity (c) of an electromagnetic wave is the distance travelled per second: c = v The velocity of an electromagnetic wave travelling through a vacuum is c = 3 x 108 m s^{-1}. The velocity of an electromagnetic wave travelling through a material is always less than that in a vacuum. The refractive index of a material is equal to $c_{vacuum}/c_{material}$.

— The energy (E) of the photons in an electromagnetic wave is related to the frequency of the wave:

$$E = h\nu = h / T = hc / \lambda = hc$$

where, h = Planks constant (6.6262 x10^{-34} J s). These expressions can be used to relate the energy of an electromagnetic wave to its frequency, period, wavelength or wave number. This relationship indicates that monochromatic radiation contains photons that all have the same energy.

The electromagnetic spectrum consists of radiation that ranges in wavelength from 10^{-12} m (high energy) to 10^{4} m (low energy). The physical principles and mathematical description of radiation across the whole of the electromagnetic spectrum is the same, however, it is convenient to divide it into a number of different regions depending on the origin of the waves, i.e., cosmic rays, gamma rays, x-rays, ultraviolet, visible, infrared, microwaves, and radio waves.

Radiation and Matter

Spectroscopic techniques utilise the fact that atoms and molecules have a discrete set of energy levels and that transitions can only occur between them. When an electromagnetic wave propagates through a material the atoms or molecules can absorb energy and move to an excited state if the photons in the wave have energies that are exactly equal to the difference between two energy levels (E = hν).

Alternatively, if an excited atom or molecule emits energy in the form of radiation the waves emitted must have energies that are exactly equal to the difference between two energy levels ($E = h\nu$).

The energy of the photons in different regions of the electromagnetic spectrum corresponds to different types of energetic transition that can occur in atoms and molecules, e.g., electronic, rotational, vibrational, translational, nuclear transitions. Electromagnetic radiation can therefore be used to probe different molecular characteristics of matter.

The difference between electronic energy levels is greater than between vibrational energy levels, which is greater than between rotational energy levels. Thus higher energy radiation (shorter wavelength) is needed to cause transitions between electronic levels than between vibrational or rotational levels. In practice, a molecule can be thought of as having a number of different electronic energy levels, with rotational and vibrational energy levels superimposed on them.

Absorption Process

Absorption is the process by which energy is transferred from an electromagnetic wave to an atom or molecule and causes it to move to an excited state. Absorption can only occur when an atom or molecule absorbs a photon of light that has an energy which exactly corresponds to the difference between two energy levels, i.e., it must be quantised. At room temperature the ground state of atoms and molecules is usually the one which is most highly populated and so transitions usually occur from the ground state to higher energy levels.

At higher temperatures, more of the higher energy levels are occupied and so, transitions between higher energy levels may also become important. If an atom or molecule is subjected to electromagnetic radiation of different wavelengths (energies) it will only absorb photons at those wavelengths which correspond to exact differences between two different energy levels within the material. A plot of the fraction of photons absorbed at a particular wavelength versus the energy of the photons at that wavelength is called an absorption spectrum. Conventionally, the axes of absorption spectra are specified in terms of easily measurable quantities: x-axis transmittance or absorbance (rather than fraction of photons absorbed); y-axis wavelength, frequency or wave number (rather than photon energy).

Emission of Radiation

Emission of radiation is the reverse of absorption, occurring when energy from an atom or molecule is released in the form of a photon of radiation. When a molecule is raised to an excited state it will only exist in this state for a very short time before relaxing back to the ground state. This is because it will always try to move to its lowest energy state. There are two important relaxation processes through which an excited molecule can dissipate its energy:

— *Non-radiative decay*: This is the most common way that an excited molecule loses its energy. Energy is dissipated in a number of small (quantised) steps due to transfer of energy from the exited molecule to surrounding molecules in the form of kinetic energy (heat). Nevertheless, the heat generated is usually so small that it has little effect on the overall temperature of the system.

— *Radiative decay*: In some cases an atom or molecule loses its energy in the form of a photon (emission). This is the case in atomic emission spectroscopy.

Sometimes both of these processes occur together. In fluorescence spectroscopy, a molecule absorbs electromagnetic radiation, which causes it to move into an excited state. It then returns to the ground state by dissipating some of its energy in the form of non-radiative decay and the rest in the form of a photon of radiation. The photon emitted is therefore of lower energy (longer wavelength) than the incident wave. Usually, an electron decays to the lowest energy level in the excited electronic state, and then returns to the ground state.

Measurement Modes

The design of an analytical instrument based on spectroscopy depends on the nature of the energetic transitions involved, the nature of the radiative process involved and the nature of the food matrix. These factors determine the wavelength (frequency) of electromagnetic radiation used, the way that the electromagnetic radiation is generated and the way that the electromagnetic radiation is detected. Some commonly used designs are highlighted below:

- *Emission*: The sample being analysed is energetically stimulated and the amount of electromagnetic radiation produced by the sample is measured at different wavelengths, e.g., atomic emission spectroscopy, NMR, fluorescence.
- *Transmission*: An electromagnetic wave generated by the analytical instrument is propagated directly through the sample and the reduction in its amplitude due to interaction with the sample is measured at different wavelengths, e.g., atomic absorption spectroscopy, infrared transmission measurements, UV-visible spectrophotometery.
- *Reflection*: An electromagnetic wave generated by the analytical instrument is reflected from the surface of the sample and the reduction in its amplitude due to interaction with the sample is measured at different wavelengths, e.g., infrared reflection measurements, colour measurements.

Spectroscopic Analysis

Quantitative Analysis

One of the most important applications of the interaction between electromagnetic radiation and matter is the determination of the concentration of certain components in foods. This application relies on there being a relationship between the amount of radiation absorbed by a material and the concentration of the components present. The power (P) of an electromagnetic wave exiting a solution is less than the power entering the solution (P_0), because solute molecules absorb some of the energy.

The amount of energy absorbed is usually expressed in terms of either the transmittance or the absorbance. The transmittance is simply the ratio of the exiting and incoming radiation: $T = P/P_0$, and is often expressed as a percentage $\%T = (P/P_0) \times 100$. Unfortunately, T or %T are not proportional to the concentration of the absorbing species and so another parameter, known as the absorbance A, has been defined which is proportional to the concentration:

$$A = -\log (P/P_0) = -\log T$$

In dilute solutions the absorbance is proportional to the concentration of the absorbing species, which is extremely convenient for quantitative analysis of concentration. The relationship between the absorbance of a solution and its concentration is known as Beer's Law:

$$A = abc$$

Here a is a constant called the absorptivity which depends on the molecular properties of the absorbing species and the wavelength of the radiation, b is the pathlength of the sample and c is the concentration of the sample.

Qualitative Analysis

Spectroscopy techniques can also be used to provide valuable information about the type, structure and environment of molecules present in food materials:

— *Atomic or Molecular Type*: Each type of atom or molecule has a unique set of energy levels and therefore a unique electromagnetic spectrum. By measuring the electromagnetic spectrum of a material and identifying the magnitude and position of the absorption or emission peaks it is often possible to determine the type of atoms or molecules present.

— *Molecular Structure*: Certain kinds of molecular groups have characteristic absorption or emission peaks in specific regions of the electromagnetic spectrum. For example, certain kinds of molecular groups give absorption peaks at specific wavelengths in an infrared spectrum, while the number, type and organisation of atoms and bonds within a molecule leads to characteristic absorption peaks in a NMR spectrum. It is therefore possible to obtain important information about the structure of molecules by measuring their electromagnetic spectra.

— *Molecular Environment*: Spectroscopy techniques can also be used to provide information about the molecular environment of atoms and molecules within a sample. The absorption or emission of energy between two energy levels within a specific atom or molecule is influenced by the presence of other atoms and molecules in their immediate vicinity. Consequently, if the molecular environment of an atom or molecule within a sample is altered, then its absorption or emission spectra may change. Spectroscopy can therefore be used to monitor physicochemical changes that result in an alteration in the molecular environment of atoms and molecules, e.g., protein unfolding, solubilisation, or aggregation.

Rheological Testing

Rheology is the science concerned with the deformation and flow of matter. Most rheological tests involve applying a force to a material and measuring its flow or change in shape. Rheology is important in a number of different areas of food science. Many of the textural properties that human beings perceive when they consume foods are largely rheological in nature, e.g., creaminess, juiciness, smoothness, brittleness, tenderness, hardness, etc. The stability and appearance of foods often depends on the rheological characteristics of their components, e.g., emulsions, spreads and pastes.

The flow of foods through pipes or the ease at which they can be packed into containers is largely determined by their rheology. The obvious importance of rheology in foods means that it is essential for food scientists to have analytical techniques to measure these properties. Instruments are needed for routine analysis in quality assurance laboratories, and for fundamental studies in Research and Development laboratories.

Fundamental studies aim to better understand the complex relationship between the overall rheological properties of foods and the type and concentration of ingredients that they contain. This type of information enables food manufacturers to optimise the ingredients and processing conditions needed to produce high quality and reliable products. Foods are compositionally and structurally complex systems that can exhibit a wide range of different rheological behaviours, ranging from low viscosity fluids to hard solids. One of the main objectives of food rheologists is to develop instrumentation and concepts that can be used to measure and describe these various types of rheological behaviour.

Despite the diversity and complexity of food systems it is possible to systematically characterise many of their rheological properties in terms of a few simple models: the ideal solid, the ideal liquid, and the ideal plastic. Complex systems can then be described by combining two or more of these simple models. In the following sections the concepts of the ideal solid, ideal liquid and ideal plastic will be introduced, as well as some of the deviations from these models that commonly occur in foods.

Solid Materials

In our everyday lives we come across solid materials that exhibit quite different rheological properties. Some may be soft, others hard; some may be brittle, others rubbery; some may break easily, others may not. Despite this range of different behaviour it is still possible to characterise the rheological properties of many solid foods in terms of a few simple concepts.

Ideal Solids

A material that exhibits ideal elastic behaviour is referred to as a Hookean solid after the scientist (Robert Hooke) who first described it. Hooke observed experimentally that there was a linear relationship between the deformation of a solid material and the magnitude of the applied force. In fact, he found that the force per unit area (or stress) was proporitional to the relative deformation (or strain). Hookes law can be summarised by the following statement:

$$\text{Stress} = \text{Modulus} \times \text{Strain}$$

Most elastic materials only obey Hookes law at small deformations. Equation 1 applies to a number of different types of deformation that a solid can experience. The actual values of the stress, strain and constant used in the equation depend on the nature of the deformation. For an isotropic and homogeneous solid there are three major types of deformation that are important: simple shear, simple compression (or elongation) and bulk compression. Each of these different types of deformation can be characterised by its own stress-strain relationship.

Simple shear:

$\text{Stress} = \tau = F/A$

$\text{Strain} = \gamma = \Delta L/L = \cos\ \phi$

Modulus = G (shear modulus)

Simple compression:

$\text{Stress} = F/A$

$\text{Strain} = \Delta L/L$

Modulus = Y (Young's modulus)

Bulk compression:

Stress = τ = F/A = Pressure, P

Strain = $\Delta V/V$

Modulus = K (Bulk modulus)

Non-Ideal Solids

Hooke's law is only strictly applicable to elastic materials at low strains, and so most fundamental rheological studies of foods have been concerned with small deformations. Nevertheless, the rheological behaviour of foods at large deformations is often more relevant to their actual use, e.g., mastication or cutting of foods. For this reason it is important to be able to systematically characterise the behaviour of solids at large deformations.

At strains just above the Hookes region the stress is no longer proportional to the strain, and therefore an apparent modulus is defined (just as an apparent viscosity is defined for non-Newtonian liquids). It is always necessary to stipulate the strain at which the apparent modulus of a material is measured. Even though the material does not obey Hookes law it still returns to its original shape once the force is removed.

Above a certain deformation, however, a solid may not return back to its original shape once the force is removed, because it either breaks or flows. A material that breaks is referred to as brittle, whereas a material that flows is referred to as plastic. The stress at which a material ruptures is often called the breaking strength. A material usually ruptures or flows because the forces that hold the atoms or molecules together are exceeded.

Liquid Foods

Liquid foods also exhibit a wide range of different rheological properties. Some have very low viscosities and flow easily, like water or milk, whilst others are very viscous, like honey or syrup. Even so, it is still possible to characterise their rheological properties using a few simple concepts.

Ideal Liquids

The ideal liquid is often referred to as a Newtonian liquid after the scientist who first described it (Sir Isaac Newton). The ideal liquid has the following characteristics: it is incompressible (its volume does not change when a force is applied to it); isotropic (its properties are the same in all directions); and structureless (it is homogeneous). The rheological properties of the ideal liquid are defined by the following equation, which encapsulates the experimental finding that the rate of shear strain is proportional to the applied shear stress τ.

$$\text{Stress} = \text{Viscosity} \times \text{Rate of Strain}$$

$$\tau = \eta d\gamma / dt$$

where, the constant of proportionality, η is called the viscosity. The viscosity arises from the friction between the liquid layers as they slide past one another. The lower the viscosity of a liquid, the less resistance between the liquid layers, and therefore the smaller the force required to cause the top plate to move with a given velocity, or the faster the top plate moves when a given force is applied.

The ideal viscous fluid differs from the ideal elastic solid because the shear stress is proportional to the rate of strain, rather than the strain. The units of shear stress τ are Nm^{-2} (or Pa), and those of shear rate are s^{-1} thus the viscosity η has units of Nsm^{-2} (or Pa s) in the S.I. system. Viscosity can also be expressed in the older c.g.s. units of Poisse, where 1Pa s = 10 Poisse. Thus the viscosity of water can be quoted as 1 mPa s, 0.001 Pa s, 0.01 Poise or 1 centipoise, depending on the units used. A number of foods exhibit ideal Newtonian behaviour under certain conditions, e.g., water, tea, coffee, oils, honey and milk.

Non-Ideal Liquids

Non-ideality may manifest itself in a number of different ways, e.g., the viscosity of a liquid may depend on the rate and/or the time over which the shear force is applied, or the fluid may exhibit some elastic as well as viscous properties.

Shear-rate dependent non-ideal behaviour

In an ideal liquid the viscosity is independent of the shear rate. In many liquid foods the viscosity varies with the shear rate, but is independent of the length of time that the food is subjected to the shear. For example, the viscosity of a liquid food may increase or decrease as the shear rate is increased, rather than staying constant as for a Newtonian liquid. In these foods the viscosity is referred to as an apparent viscosity, because it is no longer a constant.

The dependence of the apparent viscosity on shear rate, means that it is crucial to stipulate the shear rate used to carry out the measurements. The choice of shear rate to use when measuring the apparent viscosity of a non-ideal liquid is a particularly important consideration when carrying out rheological measurements in a laboratory which are supposed to mimic some process which occurs in a food naturally, e.g., flow through a pipe, the creaming of an emulsion droplet, mastication. The test in the laboratory should use a shear rate which is as close as possible to that which the food actually experiences in practice. The two most common types of shear-rate dependent non-ideal liquids are:

— *Pseudoplastic fluids*: Pseudoplastic flow is the most common type of non-ideal behaviour exhibited by liquid foods. It manifests itself as a decrease in the apparent viscosity of a fluid as the shear rate is increased, and is therefore referred to as shear thinning. Pseudoplasticity may occur for a number of different reasons, e.g., polymers may align themselves with the flow field, solvent molecules bound to a particle may be removed, or aggregated particles may break down.

— *Dilatant fluids*: Dilatant behaviour is much less common than pseudoplastic behaviour. It manifests itself as an increase in the apparent viscosity as the shear rate is increased, and is therefore sometimes referred to as shear thickening.

Time-dependent non-ideal behaviour

The apparent viscosity of the fluids described in the previous section depended only on the shear rate, and not on the length of time that the shear was applied. There are many foods whose rheological properties do depend on the duration of the applied shear. In some cases this change is reversible and the fluid will recover its original apparent viscosity if it is allowed to stand at rest for a sufficiently long period. In other cases the change brought about by shearing the sample is irreversible.

An appreciation of the time-dependency of the flow properties of foods is of great practical importance in the food industry. The duration of pumping or mixing operations, for instance, must be carefully controlled to assure that the food sample has the most appropriate apparent viscosity. If a food is mixed or pumped for too long it may become too thick or too runny and thus loose its desirable rheological properties. Time dependent non-Newtonian behaviour is classified in two different ways:

- *Thixotropic fluids*: A thixotropic fluid is one in which the apparent viscosity decreases with time when the fluid is subjected to a constant shear rate. Fluids of this type are thought to contain small particles (droplets, crystals or biopolymers) that are aggregated together by weak forces. Shearing of the material causes the aggregated particles to be disrupted and so they offer less resistance to flow and the viscosity decreases with time until a constant value is reached. This constant value may correspond to the point where the rate of structure disruption is equal to the rate of structure reformation, or where there is no more structure to be broken down. Once the shear force is removed the aggregates may reform with time as the particles collide into one another due to Brownian motion.
- *Rheopectic fluids*: In some foods, the apparent viscosity of the fluid increases with time when it is subjected to a constant shear rate. Again there may be a number of

different reasons for this. One of the most important is that shearing increases the frequency of collisions between droplets or particles in fluids that can lead to enhanced aggregation and consequently an increase in apparent viscosity.

In some fluids the time dependent rheological properties are irreversible, i.e., once the shear force is removed the system does not regain its initial rheological properties. Liquids fluids that experience permanent change are called rheodestructive. This type of behaviour might occur when aggregated particles are permanently disrupted and do not reform with time.

Plastics

Many foods exhibit a kind of rheological behaviour known as plasticity. A plastic material has elastic properties below a certain applied stress, the yield stress.

Ideal Plastic Material

The ideal plastic material is referred to as a Bingham Plastic after the scientist who first proposed this type of rheological behaviour. Two equations are needed to describe the rheological behaviour of a Bingham plastic, one below the yield stress and one above it:

$$\tau = G\gamma \qquad (\text{for } \tau < \tau_0)$$

$$\tau - \tau_0 = \eta d\gamma / dt \qquad (\text{for } \tau \geq \tau_0)$$

where G is the shear modulus, is the viscosity and 0 is the yield stress. Foods that exhibit plastic behaviour usually consist of a network of aggregated molecules or particles dispersed in a liquid matrix. For example, margarine and butter consist of a network of tiny fat crystals dispersed in a liquid oil phase. Below a certain applied stress there is a small deformation of the sample, but the weak bonds between the crystals are not disrupted.

When the critical yield stress is exceeded the weak bonds are broken and the crystals slide past one another leading to flow of the sample. Once the force is removed the flow stops. A similar type of behaviour can be observed in emulsions containing three-dimensional networks of aggregated droplets.

Non-ideal Plastic Material

Above the yield stress the fluid flow may exhibit non-Newtonian behaviour similar to that described earlier for liquids, e.g. psuedoplastic, dilatant, thixotropic, rheopectic. The material may also exhibit non-ideal elastic behaviour below the yield stress, e.g., the yield point may not be sharply defined, instead, the stress may increase dramatically, but non instantaneously, as the shear rate is increased.

Viscoelasticity

Most food materials are not pure liquids, or pure solids, but have rheological properties that are partly viscous and partly elastic. Plastic materials exhibit elastic behaviour below the yield stress, and viscous behaviour above the yield stress. In contrast, viscoelastic materials exhibit both viscous and elastic behaviour simultaneously. When a force is applied to a viscoelastic material it does not instantaneously take-up its new dimensions, it takes some finite time.

In addition, when the force is removed the material does not return instantaneously back to its non-deformed state, and it may even remain permanently deformed. Two types of experimental tests are used by food scientists to characterise the viscoelastic properties of foods: transient and dynamic measurements. Both types of tests can be carried out using simple shear, simple compression or bulk compression of foods, depending on how the instruments are designed. Since shear tests are the most commonly used in the food industry at present only these will be considered. Nevertheless, simple and bulk compression tests can also be carried out in a similar manner.

Transient Experiments

In a transient experiment a constant force is applied to a

material and the resulting strain is measured as a function of time, or vice versa.

— *Creep*: In a creep experiment a constant stress is applied to a sample and the corresponding strain is followed as a function of time. Results are expressed in terms of a parameter called the compliance J = strain/stress, because the stress remains constant. The change in strain of a material can also be measured when the stress is removed, i.e. creep recovery. Viscoelastic materials can often be characterised by a modulus and a relaxation time, which can be determined by an analysis of the strain curves with time. A distinction is usually made between a viscoelastic solid and a viscoelastic liquid. When a constant force is applied to a viscoelastic solid the creep compliance reaches a finite equilibrium value (JE) at long times. When the force is removed the compliance tends to zero. On the other hand, when a constant force is applied to a viscoelastic liquid the compliance continues to increase at a steady rate, and when the force is removed the material does not return to its initial shape.

— *Stress relaxation*: Instead of applying a constant force and measuring the change in the strain with time, it is also possible to apply a constant strain and measure the change in the stress with time. These types of experiments are referred to as stress relaxation. The same types of information can be obtained from either creep or stress relaxation experiments, and the method used usually depends on the instrument available.

Dynamic Experiments

In a dynamic experiment a sinusoidal stress is applied to a material and the resulting sinusoidal strain is measured, or vice versa. In a dynamic experiment, a sinusoidal stress is applied to a material and the resulting sinusoidal strain is measured, or vice versa. In this section, we will only consider the case where a stress is applied to the sample and the resultant strain is measured.

The applied stress is characterised by its maximum amplitude (τ_0) and its angular frequency (ω) The resulting strain has the same frequency as the applied stress, but its phase is different because of relaxation mechanisms associated with the material. Information about the viscoelastic properties of the material can therefore be obtained by measuring the maximum amplitude (γ_0) and phase shift (δ) of the strain.

The amplitude of the applied stress used in this type of test is usually so small that the material is in the linear viscoelastic region, i.e., the stress is proportional to the strain, and the properties of the material are unaffected by the experiment. The dynamic shear rheological properties of a material can be described by the complex shear modulus $G = G' + iG''$, where the parameters G' and G'' are referred to as the storage modulus and loss modulus, respectively.

This is because G' is the measure of the energy stored in the material per cycle, whereas G'' is a measure of the energy dissipated as heat (and therefore lost) per cycle. For a perfectly elastic material the stress and strain are completely in phase, and for a perfectly viscous material all the energy is lost as heat and the stress and strain are 90° out-of-phase.

The phase angle that the stress lags behind the strain is given by the symbol δ. The phase angle of a material provides a useful insight into its viscoelastic properties: $\delta = 0°$ for a perfectly elastic solid; $\delta = 90°$ for a perfectly viscous fluid; and, $0 < \delta < 90°$ for a viscoelastic material. The more elastic a material (at a particular frequency), the smaller the phase angle, and the lower the amount of energy dissipated per cycle.

Measurement of Rheological Properties

Foods are diverse and complex materials which exhibit a wide range of different rheological properties, e.g., solids, liquids, plastics and viscoelastic behaviour. Consequently, a variety of different instruments have been developed for characterising their rheological properties. Instruments vary according to the type of deformation they apply to the sample (shear, compression, elongation or some combination), the property measured, the cost, the ease of operation etc.

In many industrial applications it is necessary to have instruments which make measurements that are rapid, low-cost, simple to carry-out and reproducible, rather than giving absolute fundamental data. Thus simple empirical instruments are often used, rather than the sophisticated and expensive instruments often used in research and development. The information obtained from these instruments is difficult to relate to the fundamental rheological properties of a material because the stresses and strains applied are not easily defined. Rather than having a simple elongation, shear or compression, different types of forces may be applied simultaneously.

To compare data from different laboratories it is necessary to carefully follow standardised test procedures. These procedures may define experimental parameters such as the sample size and preparation procedure, the magnitude of the force or deformation, the design of the device used, the speed of the probe, the length of time the force is applied for and the measurement temperature. For food scientists involved in research and development it is often more important to use instruments that provide information about the fundamental rheological constants of the material being tested. These instruments are designed to apply well-defined stresses and strains to a material in a controlled manner so that stress-strain relationships can be interpreted using suitable mathematical analysis.

Rheological properties determined using these techniques can be compared with measurements made by other workers in the literature or in other laboratories. In addition, measured rheological properties can be compared with predictions made using various mathematical theories that have been developed to relate the structure and composition of materials to their fundamental rheological properties. There is an increasing trend in the food industry to use instruments that provide more fundamental data where ever possible.

Instruments can be conveniently categorised according to whether they utilise simple compression (or elongation) or shear forces. At present few instruments utilise bulk compression to analyse the rheological properties of foods.

Compression and Elongation

These types of measurements are most frequently carried out on solid or semi-solid foods that are capable of supporting their own weight. Fundamental measurements are usually carried out using instruments referred to as Universal Testing Machines. The solid sample to be analysed is placed between a fixed plate and a moving probe. The probe can have many different designs depending on the type of information required. Some of the most commonly used designs include: a flat plate, a blade, a cylindrical spike and a set of teeth! The type of probe used may also depend on whether or not the analyst is trying to mimic some actual process, e.g., chewing, biting or cutting.

The probe can be moved vertically, either upwards or downwards, at a controlled speed. The lower plate usually contains a pressure sensor that measures the force exerted on the sample when it is deformed by the probe. Thus the instrument measures both the stress and strain on the sample as it is compressed. Some of the common tests carried out using Universal Testing Machines are:

— *Stress and Strain*: The stress on a sample is measured as a function of strain. The resulting "rheogram" can be used to characterise the rheological properties of a sample. The slope of the stress versus strain relationship at small deformations is often a straight line, whose gradient is equal to the elastic modulus. At large deformations the sample may rupture and the breaking stress and strain can be determined. This type of test is used commonly to test solid samples and gels. An extension of this test is to cycle the probe upwards and downwards a number of times. The rheological properties of the food may change during each compression cycle, which may give some indication of what happens when a food is chewed in the mouth, i.e., the breakdown of food structure.

— *Stress (or Strain) and Time*: The sample is compressed to a known deformation and the relaxation of the stress with time is measured (stress relaxation). Alternatively a constant stress could be applied to the sample and the variation of strain measured with time (creep). This type of experiment is particularly useful for characterising the viscoelastic properties of food samples, e.g., relaxation times.

By using different fixtures the same instruments can be used to carry out elongation experiments. A sample is clamped at both ends, then the upper clamp is moved upwards at a controlled speed and the force required to elongate the sample is measured by the pressure sensor. Again the elastic modulus and breaking strength can be determined. Universal Testing Machines can also be adapted to perform various other types of experiments, e.g., bending or slicing.

Recently a number of more sophisticated instruments, based on dynamic rheological measurements, have been developed to characterise the rheological properties of solids, plastics and viscoelastic materials. As well as carrying out standard compression measurements, they can also be used to carry out transient or dynamic compression measurements on viscoelastic materials. These instruments are usually expensive, and are therefore only available to large food companies and some Research laboratories. Nevertheless they are extremely powerful tools for carrying out fundamental studies on food materials.

The rheological properties of a sample can be measured as a function of time or temperature, and thus processes such as gelation, aggregation, crystallisation, melting and glass transitions can be monitored. Some complications can arise when carrying out simple compression experiments. There may be friction between the compressing plates and the sample that can lead to the generation of shear as well as compression forces. For this reason it is often necessary to lubricate the sample with oil to reduce the effects of friction. In addition, the cross-sectional area of the sample may change during the course of the experiment, which would have to be taken into account when converting the measured forces into stresses. Finally, for viscoelastic materials, some stress relaxation may occur during the deformation, thus the data depends on the rate of sample deformation.

Shear Measurements

Instruments that measure shear are used to characterise the rheological properties of liquids, viscoelastic materials, plastics and solids. The instrument and test-method used depends on the nature of the sample to be analysed. Some instruments are only useful for low viscosity ideal liquids, others for solids, and others can be used for a wide range of different materials.

Some instruments are capable of measuring the viscosity over a wide range of shear rates, whereas others make the determination at a single shear rate (and are therefore only suitable for analysing Newtonian liquids). Some instruments are only capable of carrying out transient measurements, whereas more sophisticated instruments are also capable of carrying out dynamic measurements. To make accurate and reliable measurements it is important to select the most appropriate instrument and test method, and to be aware of possible sources of experimental error.

Capillary Viscometers

The simplest and most commonly used capillary viscometer is the Ostwald viscometer. This consists of a glass U-tube into which the sample to be analysed is poured. The whole arrangement is placed in a thermostated water-bath to reach the measurement temperature. The viscosity of the liquid is measured by sucking up liquid into one of the arms of the tube using a slight vacuum and then measuring the time taken for it to flow back through a capillary of known radius and length. The time t taken to travel through the capillary is related to the viscosity by the following equation:

$$t = C \times \eta / \rho$$

where, ρ is the density of the fluid, t is the measured flow time and C is a constant which depends on the precise size and dimensions of the U-tube. The higher the viscosity of the fluid, the longer it takes to flow through the tube. The simplest method for determining the viscosity of a liquid is to measure its flow time and compare it with that of a liquid of known viscosity, such as distilled water:

$$\eta_s = \eta_0 (t_s \rho_s / t_0 \rho_0)$$

where, the subscripts s and 0 refer to the sample being analysed and the reference fluid, respectively. This type of viscometer is used principally to measure the viscosity of Newtonian liquids. It is unsuitable for analysing non-Newtonian liquids because the sample does not experience a uniform and controllable shear rate. U-tubes with capillaries of various diameters are available to analyse liquids with different viscosities: the larger the diameter, the higher the viscosity of the sample that can be analysed.

Mechanical Viscometers and Dynamic Rheometers

A number of analytical instruments have been designed that can measure the shear properties of liquids, viscoelastic materials, plastics and solids. These instruments are usually computer controlled and can carry out sophisticated test procedures as a function of time, temperature, shear rate or frequency. Most of these instruments can be adapted to carry out tests using either the concentric cylinder, cone-and-plate or parallel plate arrangements discussed below.

All of these arrangements can be used to measure the viscosity of liquids, the viscoelasticity of semi-solid foods or the elasticity of solids. The instruments can be divided into two different types: constant stress instruments apply a constant torque to the sample and measure the strain or rate of strain generated, whereas constant strain instruments apply a constant strain or rate of strain and measure the torque generated in the sample. For convenience, we will just mention constant stress instruments below.

— *Concentric cylinder*: The sample is placed in the gap between two concentric cylinders. The inner cylinder is then driven at a constant torque (angular force) and the strain (angular deflection) or rate of strain (speed at which the inner cylinder rotates) is measured, depending on whether one is analysing a predominantly solid or liquid sample. For a solid, the angular deflection of the inner cylinder from its rest position is an indication of its elasticity: the larger the deflection the smaller the shear modulus. For a liquid, the viscosity of the fluid between the plates governs the speed at which the inner cylinder rotates: the faster it spins at a given torque the lower the viscosity of the liquid being analysed. The torque can be varied in a controlled manner so that the (apparent) elastic modulus or viscosity can be measured as a function of shear stress. This instrument can be used for measuring the viscosity of non-newtonian liquids, the viscoelasticy of semi-solids and the elasticity of solids.

— *Parallel Plate*: In this instrument the sample is placed between two plates: the bottom one is stationary and the top one rotates. A constant torque is applied to the upper plate, and the angular deflection or rate of strain is measured, depending on whether one is analysing a predominantly solid or liquid sample. The main problem with this type of experimental arrangement is that the shear strain varies across the sample. The shear strain in the middle of the sample is less than that at the edges. Thus parallel plate arrangements are only suitable for samples where the rheological properties are independent of shear rate, and are therefore not suitable for non-ideal liquids or solids.

— *Cone and Plate*: This is essentially the same design as the parallel plate instrument, except that a cone replaces the upper plate. The cone is specially designed to have a slight angle so that there is a constant shear strain across the sample. Thus it can be used to analyse non-ideal materials.

Any of these arrangements can be used to carry out simple viscosity measurements on fluids, by measuring the variation of shear stress with shear rate. However, some of them can also be used for more expensive applications such as the transient and dynamic rheological tests mentioned earlier. Typically the rheological properties of samples are measured as a function of time or temperature.

Empirical Techniques

Many of the techniques mentioned above are unsuitable for application in the food industry because the instrumentation is too expensive, requires highly skilled operators or measurements take too long to carry out. For this reason a large number of highly empirical techniques have been developed by food scientists.

Many of these empirical techniques have become widely accepted for analysing specific food types. Typical examples may be penotrometers to measure the hardness of fats, specially designed guillotines for analysing meat samples, devices for measuring the flow of sauces when release from a cup etc. It is difficult to analyse the data from these devices using fundamental concepts because it is difficult to define the stresses and strains involved. Nevertheless, these devices are extremely useful where rapid empirical information is required.

Some Applications

Gels: Gels are good systems for fundamental rheological studies because they are usually isotropic and homogeneous and can be prepared in many different shapes. Consequently, a huge amount of work has been carried out on characterising the rheological properties of food gels. Both simple compression and shear measurements are used routinely. Typical experiments might be:

— Prepare a solution of the protein or polysaccharide to be analysed. Place it in a dynamic rheological device which measures the shear modulus of samples. Heat or cool the sample at a controlled rate so that it gels and measure the temperature at which gelation occurs, the rigidity of the gel (shear modulus) and possibly the breaking strength of the final gel.

— Make a gel sample of standard shape and dimensions. Place the gel in a Universal Testing Machine and compress it at a known speed (typically 10 mm min^{-1}). The variation of the stress with strain is recorded. From this graph it is possible to determine the Youngs modulus of the gel and its breaking strength.

The aim of these types of study is to determine the relationship between the structure and interactions of the various ingredients in foods and the final rheological properties of the gel. This is important when developing functional ingredients that act as gelling agents in foods, or to determine the best processing conditions.

Cheese: Most cheeses are also homogeneous and isotropic materials and are therefore amenable to fundamental studies using standard compression or shear tests. It is often important to find out the relationship between the rigidity or breaking strength of a cheese and variations in its composition or the processing conditions used to manufacture it.

Thus it is possible to determine the optimal ingredients or processing conditions required to produce a high quality product. This has become increasingly important recently with the attempts of many manufacturers to develop low-fat cheeses that have properties that mimic their full-fat analogs. Attempts are often made to relate rheological measurements to sensory characteristics such as firmness, chewiness and crumbliness.

Mayonaisse: It is important that mayonnaise products have thick and creamy textures, but that they are not so viscous that they will not flow out of the bottle. In addition, it is often necessary for them to have a small yield stress so that they do not collapse under their own weight once they have been poured onto a plate or salad. The rheological properties depend on their composition, e.g., the concentration of oil droplets present, or the concentration of thickening agents.

Rheological equipment is needed to characterise the properties of mayonnaise products, and to elucidate the contribution of the various ingredients which they contain. Typically the deformation of the product may be measured as a function of shear rate in order to determine the yield stress.

Margarines and Spreads: As mentioned earlier it is important that spreadable products such as margarines and low-fat spreads retain their shape when they are removed from the refrigerator, but that they spread easily when a knife is applied. Thus they must exhibit plastic properties: i.e., have yield stresses below which they are elastic and above which they are viscous. It is usually necessary for these products to exhibit their properties over a relatively wide range of temperatures. Rheological instruments are therefore needed to characterise the properties of these systems to ensure that they do exhibit the appropriate plastic behaviour. Just as with mayonnaise the deformation of a product with increasing shear stress might be measured to determine the yield stress of a product.

Meat: Meat is a complex biological material, which is heterogeneous and non-isotropic. It is therefore difficult to carry out fundamental rheological measurements on this type of product. In addition, food scientists are often interested in properties such as the tenderness or chewiness of a meat product that are complex sensory properties, consisting of both shear and compression, and usually involving large deformations. For this reason tests on meat are often carried out using empirical instruments. For example, a device has been developed which measures the force required for a blade to slice through a piece of meat.

References

Aguilera, Jose Miguel and David W. Stanley., *Microstructural Principles of Food Processing and Engineering*, Springer, 1999.

Jango-Cohen, Judith., *The History Of Food,* Twenty-First Century Books, 2005.

Magdoff, Fred; Foster, John Bellamy; and Buttel, Frederick H., *Hungry for Profit: The Agribusiness Threat to Farmers, Food, and the Environment*, September 2000.

Regmi, Anita (ed.)., *Changing Structure of Global Food Consumption and Trade,* Market and Trade Economics Division, Economic Research Service, USDA, May 30, 2001.

9

Evaluation of Milk Quality

Milk quality control is the use of approved tests to ensure the application of approved practices, standards and regulations concerning the milk and milk products. The tests are designed to ensure that milk products meet accepted standards for Chemical Composition and Purity as well as Levels of different Microorganisms.

Testing milk and milk products for quality and monitoring that Milk Products, Processors and Marketing Agencies adhere to accepted codes of practices costs money. There must be good reasons why we have to have a quality control system for the dairy industry in Kenya. The reasons are:

— *To the Milk Producer:* The milk producer expects a fair price in accordance with the quality of milk she/he produces.

— *The Milk Processor:* The milk processor who pays the producer must assure himself/herself that the milk received for processing is of normal composition and is suitable for processing into various dairy products.

— *The Consumer:* The consumer expects to pay a fair price for milk and milk products of acceptable to excellent quality.

— *The Public and Government Agencies:* These have to ensure that the health and nutritional status of the people is protected from consumption of contaminated and substandard foodstuffs and that prices paid are fair to the milk producers, the milk processor and the final consumer.

All the above-is only possible through institution of a workable quality testing and assurance system conforms to national or internationally acceptable standards.

Milk Testing and Quality Control

Milk Sampling

Accurate sampling is the first prerequisite for fair and just quality control system. Liquid milk in cans and bulk tanks should be thoroughly mixed to disperse the milk fat before a milk sample is taken for any chemical çontrol tests. Representative samples of packed products must be taken for any investigation on quality. Plungers and dippers me used in sampling milk from milk cans.

Figure. 1: Equipment used for taking milk samples

Sampling milk for bacteriological tests require a lot of care. Dippers used must have been sterilised in an autoclave or pressure cooker for at least 15mm at 120° C before hand in order not to contaminate the sample. On the spot sterilisation may be employed using 70% Alcohol swab and flaming or scaling in hot steam or boiling water for 1 minute.

Sample Preservation

Chemical Tests

Milk samples for butterfat testing may be preserved with chemicals like Potassium dichromate (1 Tablet or ½ ml 14%

solution in a ¼ litre sample bottle is adequate.) Milk samples that have been kept cooling a refrigerator or ice-box must first be warmed in water bath at 40 °C, cooled to 20°C, mixed and a sample then taken for butterfat determination. Other preservative chemicals include Sodium acid at the rate of 0.08% and Bronopol (2-bromo-2-nitro-1, 3-propanediol) used at the rate of 0.02%.

If the laboratory cannot start work on a sample immediately after sampling, the sample must be cooled to near freezing point quickly and be kept cool till the work can start. If samples are to be taken in the field e.g. at a milk cooling centre, ice boxes with ice pecks are useful.

Labelling and Records Keeping

Samples must be clearly labelled with name of farmer or code number and records of dates, and places included in standard data sheets. Good records must be kept neat and in a dry place. It is desirable that milk producers should see their milk being tested, and the records should be made available to them if they so require.

Common Testing of Milk

Organoleptic Tests

The organoleptic test permits rapid segregation of poor quality milk at the milk receiving platform. No equipment is required, but the milk grader must have good sense of sight, smell and taste. The result of the test is obtained instantly, and the cost of the test are low. Milk which cannot be adequately judged organoleptically must be subjected to other more sensitive and objective tests.

Procedure:

— Open a can of milk.

— Immediately smell the milk.

— Observe the appearance of the milk.

— If still unable to make a clear judgement, taste the milk, but do not swallow it. Spit the milk sample into a bucket provided for that purpose or into a drain basin, flush with water.

— Look at the can lid and the milk can to check cleanliness.

Judgement:

Abnormal smell and taste may be caused by:

— Atmospheric taint.

— Physiological taints (hormonal imbalance, cows in late lactation- spontaneous rancidity).

— Bacterial taints.

— Chemical taints or discolouring.

— Advanced acidification (pH < 6.4).

Clot on Boiling (C.O.B) Test

The test is quick and simple. It is one of the old tests for too acid milk (pH<5.8) or abnormal milk. If a milk sample fails in the test, the milk must contain many acid or rennet producing microrganisms or the milk has an abnormal high percentage of proteins like colostral milk. Such milk cannot stand the heat treatment in milk processing and must therefore be rejected.

Procedure

Boil a small amount of milk in a spoon, test tube or other suitable container. If there is clotting, coagulation or precipitation, the milk has failed the test. Heavy contamination in freshly drawn milk cannot be detected, when the acidity is below 0.20-0.26% Lactic acid.

Alcohol Test

The test is quick and simple. It is based on instability of the proteins when the levels of acid and/or rennet are increased and acted upon by the alcohol. Also increased levels of albumen (colostrum milk) and salt concentrates (mastitis) results in a positive test.

Procedure

The test is done by mixing equal amounts of milk and 68% of ethanol solution in a small bottle or test tube. (68 % Ethanol solution is prepared from 68 mls 96% (absolute) alcohol and 28 mls distilled water). If the tested milk is of good quality, there will be no coagulation, clotting or precipitation, but it is necessary to look for small lumps. The first clotting due to acid development can first be seen at 0.21-0.23% Lactic acid. For routine testing 2 mls milk is mixed with 2 mls 68% alcohol.

Alcohol-Alizarin Test

The procedure for carrying out the test is the same as for alcohol test but this test is more informative. Alizarin is a colour indicator changing colour according to the acidity. The Alcohol Alizarin solution can be bought ready made or be prepared by adding 0.4 grammes alizarin powder to 1 litre of 61% alcohol solution.

Acidity Test

Bacteria that normally develop in raw milk produce more or less of lactic acid. In the acidity test the acid is neutralised with 0.1 N Sodium hydroxide and the amount of alkaline is measured. From this, the percentage of lactic acid can be calculated. Fresh milk contains in this test also "natural acidity" which is due to the natural ability to resist pH changes. The natural acidity of milk is 0.16 - 0.18%.

Apparatus

- A porcelain dish or small conical flask
- 10 ml pipette, graduated
- 1 ml pipette
- A Burette, 0.1 ml graduations
- A glass rod for stirring the milk in the dish
- A Phenophtalein indicator solution, 0.5% in 50% Alcohol
- N Sodium hydroxide solution.

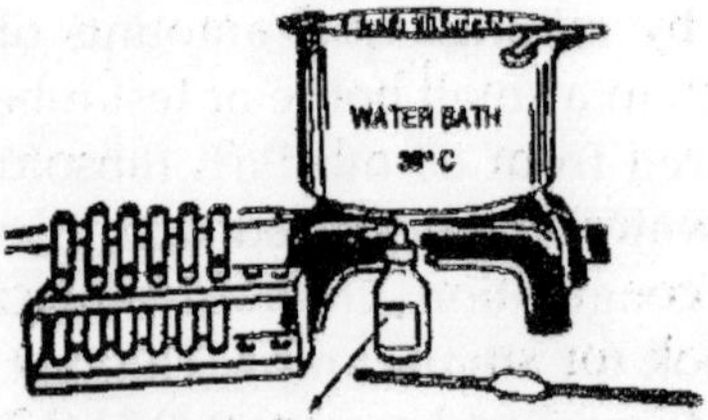

Figure. 2. Apparatus used in 10 min. Resazurin Test

Procedure

9 ml of the milk measured into the porcelain dish/conical flask, 1 ml Phenopthalein is added and then slowly from the burret, 0.1 N Sodium hydroxide under continuous mixing, until a faint pink colour appears. The number of mls of Sodium hydroxide solution divided by 10 expresses the percentage of lactic acid.

Resazurin test

Resazurin test is the most widely used test for hygiene and the potential keeping quality of raw milk. Resazurin is a dye indicator. Under specified conditions Resazurin is dissolved in distilled boiled water. The Resazurin solution can later be used to test the microbial activity in a given milk sample.

Resazurin can be carried out as:

— 10 min test.

— 1 hr test.

— 3 hr test.

The 10 min Resazurin test is useful and rapid, screening test used at the milk platform. The 1 hr test and 3 hr tests provide more accurate information about the milk quality, but after a fairy long time. They are usually carried out in the laboratory.

Apparatus and reagents:

— Resazurin tablets

— Test tubes with 10 mls mark

— 1 ml pipette or dispenser for Resazurin solution.
— Water bath thermostatically controlled
— Lovibond comparator with Resazurin disc 4/9

Procedure

The solution of Resazurin as prepared by adding one tablet to 50 mIs of distilled sterile water. Rasazurin solution must not be exposed to sunlight, and it should not be used for more than eight hours because it losses strength.

Mix the milk and with a sanitised dipper put 10 mls milk into a sterile test tube. Add one ml of Resazurin solution, stopper with a sterile stopper, mix gently the dye into the milk and mark the tube before the incubation in a water bath, place the test tube in a Lovibond comparator with Resazurin disk and compare it colourimetrically with a test tube containing 10 ml milk of the same sample, but without the dye (Blank).

Reading and Results (10 Minute Resazurin Test)

Resazurin disc No.	Colour	Grade of milk	Action
6	Blue	Excellent	Accept
5	Light blue	v. good	Accept
4	Purple	Good	Accept
3	Purple pink	Fair	Separate
2	Light pink	Poor	Separate
1	Pink	Bad	Reject
0	White	Very bad	Reject

Gerber Butterfat test

The fat content of milk and cream is the most important single factor in determining the price to be paid for milk supplied by farmers in many countries. Also, in order to calculate the correct amount of feed ration for high yielding dairy cows, it is important to know the butterfat percentage as well as well as the yield of the milk produced. Further more the butterfat percentage in the milk of individual animals must be known in many breeding programmes. Butterfat tests are also done on milk and milk products in order to make accurate adjustments of the butterfat percentage in standardised milk and milk products.

Apparatus for DF test:

- Gerber butyrameters, 0-6% or 0-8% BF
- Rubber stoppers for butyrometers
- 10.94 or 11 ml pipettes for milk
- 10 mls pipettes or dispensers for Gerber Acid
- 1 mls pipettes or dispensers for Amyl alcohol
- stands for butyrometers

Gerber water bath Reagents:

- Gerber sulphuric acid, (1.82 g/cc)
- Amyl alcohol

Treatment of samples

Fresh milk at approximately 20°C should be mixed well. Samples kept cool for some days should be warmed to 40°C, mixed gently and cooled to 20°C before the testing.

Procedure

Add 10 mIs sulphuric acid to the butyrometer followed by 10.94 or 11 mls of well mixed milk. Avoid wetting of the neck of the butyrometer. Next add 1 ml of Amyl alcohol, insert stopper and shake the butyrometer carefully until the curd dissolves and no white particles can be seen. Place the butyrometer in the water bath at 65°C and keep it there until a set is ready for centrifuging. The butyrometer must be placed in the centrifuge with the stem (scale) pointing towards the centre of the centrifuge.

Spin for 5 min. at ll00 rpm. Remove the butyrometers from the centrifuge. Put the butyrometers in a water bath maintained at 65°C for 3 min. before taking the reading. The fat column should be read from the lowest point of the meniscus of the interface of the acid-fat to the 0-mark of the scale and read the butterfat percentage. The butyrometers should be emptied into a special container for the very corrosive liquid of acid-milk, and the butyrometers should be washed in warm water and dried before the next use.

Appearance of the Test

— The colour of the fat column should be straw yellow.

— The ends of the fat column should be clearly and sharply defined.

— The fat column should be free from specks and sediment.

— The water just below the fat column should be perfectly clear.

— The fat should be within the graduation.

Problems in Test Results

Curdy tests:

— Too lightly coloured or curdy fat column can be due to:

— Temperature at milk or acid or both too low.

— Acid too weak.

— Insufficient acid.

— Milk and acid not mixed thoroughly.

Charred tests:

— Darkened fat column containing black speck at the base is due to:

— Temperature of milk-acid mixture too high.

— Acid too strong.

— Milk and acid mixed too slowly.

— Too much acid used.

— Acid dropped through the milk.

Lactometer Test

Addition of water to milk can be a big problem where we have unfaithful farm workers, milk transporters and greedy milk hawkers. A few farmers may also fall victim of this illegal practice. Any buyer of milk should therefore assure himself herself that the milk he/she purchases is wholesome and has not been adulterated. Milk has a specific gravity.

When its adultered with water or other materials are added or both misdeeds are committed, the density of milk change from its normal value to abnormal. The lactometer test is designed to detect the change in density of such adulterated milk. Carried out together with the Gerber butterfat test, it enables the milk processor to calculate the milk total solids (% TS) and solids not fat (SNF). In normal milk SNF should not be below 8.5% according to Kenya Standards (KBS No 05-10:-1976).

Procedure

Mix the milk sample gently and pour it gently into a measuring cylinder (300-500). Let the Lactometer sink slowly into the milk. Read and record the last Lactometer degree (°L) just above the surface of the milk. If the temperature of the milk is different from the calibration temperature (Calibration temperature may be=20 0C) of the lactometer, calculate the temperature correction. For each °C above the calibration temperature add 0.2° L; for each °C below calibration temperature subtract 0.2 °L from the recorded lactometer reading.

For the calculations, use lactometer degrees, and for the conversion to density write 1.0 in front of the true lactometer reading, i.e. 1.030 g/ml. Clever people may try to adulterate milk in such a way that the lactometer cannot show the adulteration. But look to see if there is an unusual sediment from the milk at the bottom of the milk can and taste to find out if the milk is too sweet or salty to be normal.

Samples of milk from individual cows often have lactometer reading outside the range of average milk, while samples of milk from herds should have readings hear the average milk, but wrong feeding, may result in low readings. Kenyan standards expects milk to have specific gravity of 1.026 -1.032 g/ml which implies a Lactometer reading range of 26.0 -32.0 °L. If the reading is consistently lower than expected and the milk supplier disputes any wrong doing arrange to take a genuine sample from the supplier.

Freezing Point Determination

The freezing point of milk is regarded to be the most constant of all measurable properties of milk. A small adulteration of milk with water will cause a detectable elevation of the freezing point of milk from its normal values of -0.54°C. Since the test is accurate and sensitive to added water in milk, it is used to detect whether milk is of normal composition and adulterated.

Inhibitor Test

Milk collected from producers may contain drugs and/or pesticides residues. These when present in significant amounts in milk may inhibit the growth of lactic acid bacteria used in the manufacture of fermented milk such as Mala, cheese and Yoghurt, besides being a health hazard.

Principle of the method: The suspected milk sample is subjected to a fermentation test with starter culture and the acidity checked after three hours. The values of the titratable acidity obtained is compared with titratable acidity of a similarly treated sample which is free from any inhibitory substances.

Materials:

- test tubes
- Starter culture
- 1 ml pipette
- water bath
- material for determination of titratable acidity

Procedure:

- Three test tubes are filled with 10 ml of sample to be tested and three test tubes filled with normal milk.
- All tubes are heated to 90 0C by putting them in boiling water for 3 - 5 minutes.

- After cooling to optimum temperature of the starter culture (30,37, or 42°C), 1 ml of starter culture is added to each test tube, mixed and incubated for 3 hours.
- After each hour, one test tube is from the test sample and the control sample is determined.

Assessment of results:

- If acid production in suspected sample is the same as the normal sample, then the suspect sample does not contain any inhibitory substances;
- If acid production as suspect sample is less than in the normal milk sample, then, the suspect sample contains antibiotics or other inhibitory substances.

Quality Control of Pasteurised Milk

When milk is pasteurised at 63°C for 30 min in batch pasteuriser or 72°C for 15 seconds in heat exchanger, continuous flow pasteurisers, All Pathogenic Bacteria are Destroyed, there by rendering milk safe for human consumption. Simultaneously various enzymes present in milk, and which might affect its flavour, are destroyed.

In order to determine whether or not milk has been adequately pasteurised, one of the enzymes normally present in milk phosphatase, is measured. A negative phosphatase result indicates that the enzyme and any pathogenic bacteria have been destroyed during pasteursation. If it is positive, it means the pasteurisation process was inadequate and the milk may not be safe for human consumption and will have a short shelf life.

- Test tubes
- 5 mls pipettes
- 1 ml pipettes
- 100 ml volumetric flask
- 500 ml volumetric flask
- water bath at 37°C

Reagents

- *Buffer Solution:* Is mixed by 0.75g anhydrous sodium carbonate and 1.75g Sodium bicarbonate in 500 ml distilled water.
- *Buffer-substrate Solution:* Place 0.15 g of di-sodium paranitrophenylphosphate (the substrate)into a clean 100ml measuring cylinder. Add the buffer solution to make to 100 ml mark. Store this buffer-substrate solution in a refrigerator and protected against light. It should not be used after one week. Prepare a fresh stock.

Procedure

Pipette 5 mls buffer-substrate solution into a test tube, stopper and warm the solution in the water bath at 37°C. Add to the test tube 1ml of the milk to be tested, stopper and mix well and place in water bath at 37°C. Prepare a blank sample from boiled milk of the same type as that undergoing the test. Incubate both the test samples and the blank sample at 37°C for 2hrs. After incubation, remove the tubes and mix them thoroughly. Place one sample against the blank in a Lovibond comparator "All Purposes" using A.P.T.W. disc and rotate the disc until the colour of the test sample is matched and read the disc number.

Interpretation:

Disc Reading after 2 hrs. incubation at 37°C	Remarks
0-10	Properly pasteurised
10-18	Slightlyunder pasteurised
18-42	Under Pasteurised
> 42	Not Pasteurised

Production of Quality Milk

History records cows being milked as far back as 9000 B.C. In Florida, cows have been milked since the settlement of St.

Augustine in 1565. However, it was not until colonial times in 1611 that dairy cows arrived in Jamestown. From that time until 1850, dairying changed very little. Most cattle were dual purpose (used for dairy and beef purposes) and were kept to satisfy family food needs. Milk and dairy products were in short supply and for the most part unavailable to those not living on or near the farm.

Milk production was seasonal, creating periods of excess as well as deficiency in the family milk supply. Stabilisation of these production fluctuations by storage and/or further processing into butter, cheese, or other milk products was precluded by the lack of refrigeration. Consequently, marketing of milk, butter, and cheese was limited to towns which could be reached by horse-drawn wagons. Over the years modern technology has rectified these problems and today a wide array of safe, wholesome dairy products are available to people throughout the developed world.

Production of quality milk is the concern of:

- consumers of dairy products
- retail distributors (super markets)
- milk and milk product processors
- dairy cooperatives
- state regulatory departments
- veterinarians, and
- dairymen.

From the list it's obvious that very few of us are left out. Whether we derive a living from the dairy industry through employment or otherwise, most of us are at the very least consumers of dairy products.

Milk Composition and Nutritional Value

The Code of Federal Regulations, Title 21, Section 131.110 provides the following definition of milk:

> "Milk is the lacteal secretion, practically free from colostrum, obtained by the complete milking of one or more healthy cows."

Nearly 12% of the American household's total food expenditure is for dairy products. Milk and milk products alone provide 10% of the total available calories in the United States food supply, and in addition, represent one of the best natural sources of essential amino acids for human nutrition. These nutritional attributes of milk have long made it a mainstay particularly in the diet of growing children. There are estimated to be some 8 to 10,000 different types of milk products available thus making it an exceptionally versatile raw product.

Milk is composed of water, fat, protein, lactose and minerals (ash). The concentration of these components will vary between cows and breeds. Total milk solids refers specifically to fat, protein, lactose and minerals. This is to be differentiated from solids-not-fat milk (SNF), a frequently used term which describes the total solids content minus fat. SNF milk is known to most people as "skim milk". The nutritional as well as economic value of milk is directly associated with its solids content. The higher the solids content the better its nutritional value and the greater the milk product yields.

Flavour and Odour Characteristics

Consumer acceptance is greatly affected by flavour. There are several factors which may produce off-flavours and/or odours in milk. Some of the more common causes of flavour and odour problems are:

- Feed and weed flavours
  - wild onion or garlic
  - strong flavoured feedstuffs such as alfalfa silage
- Cow-barny flavours - which result when milk is obtained from unclean or poorly ventilated environments, improperly cleaned or sanitised milking equipment
- Rancid flavours - presence of free fatty acids (FFA)
  - due to excessive agitation of milk during collection or transport

— breakdown of the milk fat component by proteolytic and lipolytic enzymes present in raw milk

— Malty flavours, high acid flavours
  — bacterial contamination

— Oxidised flavours
  — exposure of milk to sunlight
  — contact of milk with oxidising agents such as rust, copper, and chlorine

— Foreign flavours
  — fly sprays, medications, etc.

Process of Delivering Milk

A multitude of events take place in the process of delivering milk from the farm to the dinner table and all are designed to provide the consumer with a wholesome, nutritious and safe product. The production of quality milk and milk products begins on the farm and continues through further handling, processing and distribution.

Milk processing has three primary objectives:

— destruction of human pathogens through pasteurisation

— keeping the quality of the product without significant loss of flavour, appearance, physical and nutritive properties, and

— selective control of organisms which may produce unsatisfactory products

Milk processing plant procedures seek to:

— prevent further bacterial contamination of raw materials

— reduce bacterial numbers in milk

— protect the finished product from recontamination through careful handling, proper packaging and storage

Pasteurisation is the means whereby raw milk is rendered safe for human consumption. It is the process of heating milk to a sufficient temperature for a sufficient length of time to make it free of pathogens, however, not totally free of bacteria.

Bacteria in Milk

As stated earlier, certain organisms are capable of surviving pasteurisation and/or refrigeration processes. These bacteria are an important concern because they reduce product shelf-life. Those of major significance are:

Thermoduric bacteria

- common in raw milk
- they survive pasteurisation and include
  - Enterococci
  - Micrococci
  - Brevibacterium
  - Lactobacilli

Psychrotropic bacteria

- common dairy product contaminants
- these grow at refrigeration temperatures
- they do not survive pasteurisation
- can produce off-flavours

spore-formers

- common contaminants
- survive pasteurisation
  - Clostridial spp.
  - Bacillus spp.

The primary source of these bacteria is the environment: air, dust, dirty equipment, operators, etc. Therefore, proper cleaning and sanitising procedures are necessary for quality control. Grade A milk quality standards allow a maximum of 100,000 bacteria/ml. in raw bulk milk. Chronic offenders of these limits risk losing their license to sell milk to the Grade A market. Most dairies are able to maintain bacteria counts between 5 to 10,000 per ml. When high counts become a problem it is generally due to one or more of the following:

- — improper cleaning of milking equipment (the most common cause of high bacteria counts in milk)
- — improper cooling of milk
- — occasionally, a herd experiencing a high prevalence of infection due to Strep ag. or Staph sp.

Role of Somatic Cells in Milk

Somatic cell counts represent another important milk quality parameter. Milk with high somatic cell concentrations reduces cheese yields due to the lowered casein content. In addition, high cell count milk generally contains increased amounts of proteolytic and lipolytic enzymes (lipase). These presence of these enzymes in milk increases the potential for off- flavours and odours.

Somatic means body and thus a somatic cell is a body cell. There are three types of somatic cells typically found in milk: epithelial cells, macrophages, and polymorphonuclear leukocytes (PMN). Cell types found in milk obtained from non-infected glands are predominantly epithelial cells and macrophages. Milk from infected glands, however, generally contains high concentrations of PMN's with little or no increase in other cell types.

Consequently, somatic cell counts are an important indicator of udder health. Somatic cell counts are made available to dairymen from a variety of sources including milk quality laboratories operated by state and local regulatory departments, dairy cooperatives, DHIA-SCC programme, and veterinary diagnostic laboratories.

Mastitis causes a shift in the composition of milk. In addition to lowered amounts of casein, lactose and fat levels are decreased particularly in milk with somatic cell counts in excess of 2 million. Because the bacterial quality and somatic cell content of raw milk are important to product shelf-life, flavour and yields (particularly cheese), milk processors strive to obtain the highest quality raw product possible from their producers.

Antibiotic Residues in Milk

Antibiotic residues pose a significant public health threat. Consequently, milk in Florida is routinely monitored by dairy cooperatives and the Florida Department of Agriculture and Consumer Services, Division of Dairy Industry. The official test in current use is the Bacillus stearothermophilus disc assay. It is particularly sensitive for penicillin but can detect other inhibitors as well. The vast majority of antibiotic residues in milk occur by accident. Dairymen can avoid residue problems by:

- — properly identifying treated cows
- — informing milkers of the need to withhold and the method for withholding milk
- — keeping an accurate record of dates and times of treatment
- — following label directions and veterinarians' advice for withholding times
- — having milk tested from suspect cows if uncertain about treatment or withholding time
- — having tank milk tested when it is suspected of having milk containing antibiotic residue
- — isolate purchased cows and test their milk for residue prior to their entry into the milking herd.

Dairy Cooperatives

Dairy cooperatives are organised by dairymen for the purpose of marketing milk. Thus, instead of buying milk direct from farms milk processors buy their raw product from the dairy cooperative. Dairymen are in turn paid by the cooperative for the milk they produce. Florida's dairies supply approximately 80% of the state's marketing demands. Therefore, at certain times of the year some milk must be imported to satisfy processing needs.

During other times of the year, milk production is in excess of market demand and milk must be exported out-of-state. These daily marketing difficulties could be both expensive and time-consuming problems for dairymen. Consequently, the majority of producers belong to milk marketing cooperatives. Cooperatives serve the dairymen by promoting dairy products, providing an effective lobby for political concerns, and informing members on a variety of dairy industry issues such as water quality, waste management, and milk pricing.

Different Animal Milk Detection in Dairy Products

The adulteration of food products is a significant problem in the food production. This is how fraudulent producers try to cheat consumers and authorities. The adulteration affects all commodities in the food processing. Most frequently, such products are adulterated that are produced in big quantities and further, the expensive products whose adulteration brings a profit.

Detection of Adulteration

First of all, it is necessary to familiarise with the composition of individual milks from different species and to find the suitable marker components for the detection of adulteration with other species. The determination of fat, crude protein, lactose, ashes, and total dry matter in cow, ewe and goat buffalo milk and colostrum was dealt with by Hadjipanayiotou. The highest contents of fat, crude protein, ashes, and total dry matter were found in ewe milk followed by goat milk, the lowest contents of these components having been discovered in cow milk.

We also obtained the basic composition of buffalo milk. Table 1 presents the milk yields of individual animal species. It is apparent that the yield of cow milk is significantly higher than those of other species. It also follows from Table 1 that the highest content of non-protein nitrogen (NPN) can be found in goat milk and the lowest one in cow milk.

Table 1. Output of cow, ewe, goat and bu.alo milks and their composition

Component	Bovine	Ewe	Goat
Number of animals	44	30	30
Milk output (kg/day)	21.09 ± 7.86	2.45 ± 0.549	3.49±.504
Fat (g/kg)	38.0	55.2	45.2
Crude protein (N × 6.38) (g/kg)	33.0	58.0	41.1
NPN (g/kg)	2.18	2.70	2.91
NPN (expressed in % of crude protein)	7.43	4.66	7.13

It follows from the literature that the detection of adulteration by the substitution of one milk type for another one is made by protein analysis. This problem is a very complicated one as it is necessary to take into account that the composition of milk and milk proteins is very variable, both between individual types of milk and within one type. It depends on the breed or on the lactation level.

The quantitative determination of milk proteins is complicated by the existence of genetic and nongenetic polymorphism, and by the technological treatment and processing of milk. Thermal denaturation or proteolysis, that is common with the manufacturing of many milk products, incurs a risk of complex formation, the formation of insoluble new compounds, smaller peptides and amino acids whose analysis is fairly complicated. The information on the occurrence and quantity of individual proteins or derived compounds is, for the reasons mentioned above, very important for the estimation of processing, quality, and adulteration. Protein content in individual milks and their abundance in casein and whey fractions is shown in Table 2.

Table 2. Protein composition of bovine, ewe, goat and bu.alo milk

Component (g/100 g)	Bovine	Ewe	Goat	Bu.alo
Proteins	3.2	4.6	3.2	4.6
Caseins	2.6	3.9	2.6	4.5
Whey proteins	0.6	0.7	0.6	0.5

Many studies were published on bovine casein. Its composition in both raw and processed milk is well known. However, few studies deal with the composition of milk casein in other types of milk (e.g. goat, ewe or buffalo). Jensen stated that caseins make 82, 87, 80, and 77% of proteins in ewe, buffalo, bovine and goat milk, respectively. Casein micelles consist of four caseins: aS1-, α_{S2}-, β- and k-caseins that occur in cow milk in the ratio of 39:10:36:13.

In milk are also present the products of proteolysis of all four primary caseins. γ-Casein and some proteoso-peptone compounds are fragments of β-casein, originating from the action of plasmin, the endogenous alkaline milk protease. λ-Caseins are presumably fragments of α_{S1}-caseins, having also originated through plasmin cleavage. Glycomacropeptides and para-k-caseins are fragments of k-caseins emerging as a result of chymosin action.

Whey proteins contain proteins soluble at pH 4.6 and 20°C. To these proteins belong β-lactoglobulin (β-Lg), α-lactoalbumin (α-La), immunoglobulins (IgG, IgA, IgM) and serum albumin (BSA). The following minor proteins are also present: lactoferrin, lactoperoxidase, enzymes, protein compounds of milk fat globule membrane (MFGM), proteosopeptone compounds, and glycomacropeptides. Well known are the primary sequences α-La, β-Lg and BSA. In bovine milk, β-Lg and α-La occur approximately at a ratio of 3:1. β-Lg is the main whey protein in all types of milk studied. Its highest abundance was found in ewe milk, the lowest one in goat milk. In ewe milk, immunoglobulins are contained in significant amounts; after β-Lg they represent the second largest fraction. In bovine and goat milks, the second most represented whey fraction is α-La.

The discovery of two variants of β-lactoglobulin in cow milk by Aschaffenburg and Drewry in 1955 generated considerable interest in the research of milk proteins. The polymorphism of milk proteins is caused either by the substitution of amino acids or by their deletion. All caseins and main whey proteins show genetic polymorphism that can affect milk composition and some parameters of milk processing. For this reason, genetic variants of milk proteins are considered to be potential selection criteria in cattle husbandry. The identification of the genetic variant of milk protein also enables the determination of adulteration with various animal milk types.

Analysis of Milk Proteins and their Genetic Variants

For the detection of adulteration of one type of milk with another one, namely the following methods are used: electrophoresis, isoelectric focusing (IEF), capillary electrophoresis (CE), reverse phase high-performance liquid chromatography (RP HPLC) and ion exchange high-performance liquid chromatography (IE HPLC), hydrophobic interactive chromatography (HIC), immunochemical methods (ELISA), PCR techniques, and mass spectrometry.

Electro-migration Methods

Electrophoresis

Electrophoresis plays a significant role in the research of milk proteins and genetic variants of main milk protein components. The classification of caseins was carried out by electrophoretic analysis; minor casein components Y_1, Y_2, Y_3 and para-κ caseins were detected.

Polyacrylamide gel electrophoresis was used for the analysis of milk proteins. This employs the separation of individual molecules both according to their electric charge and size. In PAGE, the proteins have a negative or a positive charge, depending on the buffer conditions. They migrate with a speed based on their charge and size. The advantage of this technique lies in the fact that the individual groups of milk proteins are well separated and that the genetic protein variants and the levels of their phosphorylation can be detected.

Tamine et al. proved the addition of bovine milk (25% and more) into the goat Kish product by Page technique. The analysis was based on the mobility of bovine α_{S1}-casein. This technique was also used for the identification of bovine milk in ewe yoghurt. The marker component was bovine para-k-casein and 1% addition of bovine milk was successfully detected.

For the detection of adulteration also SDS and urea electrophoresis can also be used. Urea and anion solvent sodium dodecyl sulphate (SDS) possess the ability of dissolving various types of proteins and of decomposing polymer proteins to polypeptide components. SDS binds to individual protein molecules and gives them a strong negative charge thus removing the differences in the total charge. The electrophoretic separation takes place only on the basis of their molecular weights. On the contrary, with the application of urea, proteins are separated according to their charge.

Tamine et al. used for the cow milk detection in goat Kish product besides the above mentioned PAGE also SDS electrophoresis. A good separation was achieved of k-, β-, α-caseins of cow and goat milk. Nevertheless, the detection of adulterants was not successful due to the extensive proteolysis of the sample. By urea-PAGE electrophoresis, the presence of 5% and more of bovine milk in ewe and goat milk was detected. The detection was based on the analysis of bovine α_S-casein.

Isoelectric Focusing (IEF)

The protein separation according to their isoelectric points is especially suitable for the analysis of caseins that form many genetic variants. For example, by using IEF instead of PAGE electrophoresis for the analysis of the genetic variants of bovine β-casein, the procedure is significantly simplified. If for the same determination PAGE electrophoresis is to be used, it has to apply both alkali and acid PAGE for the differentiation of A variants from B, C and D, and A1, A2, and A3 variants.

IEF is an EU reference method for the determination of the presence of cow milk and caseinate in cheeses made from ewe, goat or buffalo milk or their blends. It is based on the identification of γ-caseins after plasminolysis. The method is suitable for the sensitive and specific detection of raw and thermally processed cow milk and caseinate in fresh and ripened cheeses made from ewe, goat or buffalo milk or their blends.

This detection is based on γ_2- and γ_3-caseins determination. Their isoelectric points lie between pH 6.5 and 7.5. By using two milk reference standards (with 0 and 1% of cow milk), the samples positive for the presence of cow milk can be detected. In the case, that the amounts of bovine γ_2-casein and γ_3-casein are equal or greater that their amounts in 1% standard, the presence of cow milk is confirmed. The method allows the detection of 0.5% addition. The disadvantage is that it is impossible to determine the adulteration with goat milk in ewe milk and vice versa. This may be changed by selecting an appropriate marker analyte, e.g. para-k-casein.

Capillary Electrophoresis (CE)

Capillary electrophoresis is a modification of electrophoresis that is carried out as a carrier-free electrophoresis via free capillary. CE is a quickly developing technique that enables a rapid casein and whey protein separation with a high resolution and good quantification. The use of CE resulted in the development of expedient and automated analyses with a very high resolution and with the demand for only very small amounts of samples and buffers. De Jong et al. carried out a complete analysis of caseins and whey proteins in bovine, ewe and goat milk by capillary electrophoresis. Based on this analysis, he succeeded in identifying adulterants (starting from 1%) of bovine, ewe and goat milk in all milk blends.

The employment of capillary zone electrophoresis in the detection of adulterants of cow milk in ewe and goat milk is described in the paper of Cattaneo et al. α_{S1}-Casein fraction was the marker component for the detection of cow milk and a successful detection was carried out of the addition of cow milk to ewe or goat milk starting from 8%. Lee et al. used the detection of α_{S1}-casein fraction with this technique. The authors succeeded in improving the detection and they were able to determine 1% of adulterants in raw and reconstituted milk.

Whey fraction was also studied for the detection of cow milk presence in goat milk and cheeses. As suitable marker analytes, caprine α-lactalbumine and bovine β-lactoglobuline A were determined. Minimal detectable amount of cow milk was 2% in milk mixtures and 4% in cheeses.

An unusual method using CE analysis of ethanolwater protein fractions with isoelectric iminodiacetic acid buffer was also described as a possibility to fast identify and quantify cow milk adulterants in goat and ewe cheeses. The authors declare that the amount of cow milk in goat and ewe cheese can be estimated with the relative standard deviation of 6-7%, based on electroforegrams and statistical PLS (partial last squares) multivariable regression.

References

Amigo L., *et al.*, "Genetic polymorphism of ovine milk proteins: its influence on technological properties of milk" - a review, *International Daairy Journal*, 10: 135–149, 2000.

Bania J., *et al.*, "Application of polymerase chain reaction for detection of goats' milk adulteration by milk of cow", *Journal of Dairy Research*, 68: 333–336, 2001.

Cattaneo T.M.P., *et al.*, "Analysis of cow, goat and ewe milk mixtures by capillary zone electrophoresis (CZE): preliminary approach", *Milchwissenschaft, 51:* 616–619, 1996.

10

Quality Management in Fish Processing

Quality management system should be implemented in all companies processing seafood, in vessels where processing is done and at fish auctions to fulfil all requirements regarding quality and safety. It is also known that legislation and inspection authorities recommend or require adopting the Hazard Analysis Critical Control Point (HACCP) system in production and include it as part of the quality system.

The need for HACCP in seafood industries is due to growing awareness of consumers and the fear of food illnesses. This has lead to enforced hygienic and sanitary regulations to ensure wholesome and safe food products. The growing trend in international trade for worldwide equivalence of food products is also to be considered for all producers and sellers. To guarantee safe and wholesome fishery products, a seafood processing company must have an appropriate quality system operating effectively.

The quality system includes Good Manufacturing Practices (GMP), Standard Sanitation Operational Procedures (SSOP) and a well documented HACCP plan. The HACCP system is the basis of the regulations on fish inspection adopted by the European Economic Community (EEC), USA, Canada and a number of developing countries. The quality system based on HACCP is widely used and internationally recognised by Codex Alimentarius, which recommends its adoption.

The Food and Drug Administration (FDA) regulations require fish processors to implement HACCP systems. The European Union (EU) recommends the same in its Council Directive 91/493. The maintenance of the HACCP system is as important as its implementation. Verification procedures and record keeping are strong elements of those activities. It is unlikely that the products produced, the process, the environment, likely hasards or the people in the plant will remain unchanged over time. Recording all the parameters concerning catching, handling, processing and quality is an essential tool to verify how the system is working.

Record keeping also offers product traceability in the whole chain, from catch until the final product is delivered to the consumer, so procedures for product identification and traceability during all stages should be established. Traceability is becoming an important issue for producers and sellers in all countries, not only in the EU. In Iceland the Directorate of Fisheries (DOF) through its Surveillance Department is the competent authority responsible for enforcing laws and regulations regarding handling, processing and distribution of marine products. It focuses on ensuring that fishery products are processed under satisfactory hygienic conditions and that consumers can rely on their wholesomeness and safety.

The department issues processing licenses to processors of fishery products and operating permits to fish markets and fishing vessels, provided that they meet the requirements concerning appropriate facilities, equipment, sanitation procedures and a documented check system based on HACCP. The Surveillance Department issues health certificates for exported fishery products. Developing countries are playing an important role worldwide by exporting their products to Europe and other markets, characterised by increasing demand, both on quantity and high quality products.

To be competitive, these countries also need to adopt and implement quality management systems based on HACCP and in the near future, be able to have an effective system for traceability of the products they are exporting. In the last decades Cuba has been exporting seafood products to different countries in Europe, Asia and Canada. Main products are processed lobster and shrimp with profits from exports in 1999 contributing nearly USD 200 million to the country's economy. Forty fishery establishments are engaged in fish and shellfish processing, of which 14 are certified for export.

In 1992, the Cuban fisheries industry began HACCP implementation to meet the requirements of its major export markets and to put inspection and quality control activities in line with those taking place worldwide. According to Resolution No 344/1996 of the Minister of Fisheries Industry it is mandatory for seafood processors and trading companies to put a quality system based on HACCP into operation. All the factories producing for export and most of those producing for the domestic market in Cuba have an appropriate quality system. However, problems have been observed when assessing HACCP. They include different approaches to identifying Critical Control Points (CCP), defining critical limits, and badly documented monitoring procedures.

Lack of floor staff training in applying corrective actions in case of deviations of critical limits has also been observed, as well as corrective actions that are not appropriate to eliminate causes giving rise to deviations. A field study to investigate the quality system, the traceability of products and quality monitoring in an Icelandic company is a practical experience, which will be helpful in future work in Cuba. This experience will be valuable and useful to promote the improvement of quality systems in the fisheries sector in Cuba, and especially in product traceability issues.

HACCP is a management system in which food safety is addressed through the analysis and control of biological, chemical and physical hazards from the raw material, to processing, distribution and consumption of the finished product. It was developed nearly 30 years ago by the Pillsbury Company working together with the National Atmospheric and Space Agency (NASA) in USA, with the objective of finding a method to provide safe food for astronauts.

The system focuses on preventing hazards that could cause food-borne illnesses, by applying controls to the production line, from raw material to the finished products. HACCP is a tool to assess hazards and establish control systems that focus on prevention rather than relying on end-product testing. Implementation of HACCP enhances food safety and promotes trade by increasing confidence in safe foods.

Before the application of HACCP principles the following tasks should be carried out:

— Assembling of the HACCP team, involving experts from production, quality assurance, engineering and product development areas. It could also be useful to have representatives from other areas.

— Product description and intended use. Includes principal raw materials, process technologies used, storage conditions and shelf life. This is particularly important if the product is intended for especially vulnerable groups of the population, such as infants or ill people.

— Process flow diagram. It should cover all the steps in the process, from raw material through to distribution. Such flow chart is the basis of the hazard analysis.

— On-site verification of the flow diagram. The HACCP team should check the operation against the flow diagram during all stages and hours of operation and make the amendments where appropriate.

The seven principles of HACCP are the following:

— Conduct a hazard analysis. Potential hazards associated with food and preventive measures to control those hazards.

— Identify Critical Control Points (CCP). A CCP is an operation (practice, procedure, process or location) at which a preventive or control measure will eliminate, prevent or minimise one or several hazards.

- — Establish critical limits for each preventive measure at a CCP.
- — Establish CCP monitoring procedures. Include what is to be checked, when, how and by whom.
- — Establish what corrective actions are to be taken when monitoring shows that a critical limit has been exceeded.
- — Establish verification procedures to verify that the HACCP system is working correctly.
- — Establish an effective record keeping system to document the HACCP system.

Records need to be kept as evidence that the system has been working correctly. They are useful for trend analysis, which can be used for monitoring and making the system more effective. HACCP is not a system that stands alone, it is supported by other programmes known as prerequisites, such as Good Manufacturing Practices (GMP) and Standard Sanitation Operational Procedures (SSOP). Prerequisite programmes provide the basic environmental and operating conditions that are necessary for the production of safe, wholesome food.

The Codex Alimentarius General Principles of Food Hygiene describe the basic conditions and practices expected for food intended for international trade. In addition to the requirements specified in regulations, industry often adopts procedures specific to their operations. Prerequisite programmes may include facilities, supplier control, specifications of raw materials, ingredients, packaging materials and products, equipment, cleaning and sanitation, personal hygiene, training, traceability and recall procedures, pest control, etc.

Traceability in the Fish Industry

Traceability is defined as the ability to trace the history, application or location of that which is under consideration. In terms of products it relates to the origin of materials and parts, the processing history, and the distribution of the product after delivery. In other words traceability means the ability to trace and follow a food through all stages of production and distribution.

Two types of traceability can be identified: internal and chain traceability. Internal is within one company and relates to data about raw materials and processes to the final product before it is delivered. Chain traceability is focused on the information about the product from one link in the chain to the next, it describes what data are transmitted and received, and how. Chain traceability is between companies and countries and depends on the presence of internal traceability in each link. The public confidence in food safety has been damaged by recent food scares associated with beef because of mouth and foot disease and BSE—mad cow disease, in cattle, dioxin in fish meal and other. This is driving the industry and government agencies to improve controls at all stages in the food chain.

Traceability is then needed to meet food safety requirements, especially in case of product recall, for commercial reasons to ensure supply chain standards and because it is required by legislation relating to labelling, animal health and welfare, fish marketing, fisheries control and product liability and safety. The EU Fisheries Control Regulations demand a specific traceability system from the fishing grounds to the processors. The EU Fish Marketing Regulations demand that from 2002, much of the fish at retail sale will have to be labelled with its area of origin.

The proposed revision of the EU General Product Safety Directive requires full traceability by 2003, including product recall systems. The EU Food Law, which is now under revision, requires full traceability by 2004. The proposed new Regulation on the General Principles and Requirements of Food Law lays down the general food safety requirements. Regarding traceability, it establishes the need for traceability at all stages of production and distribution. It is proposed that food and feed business operators must identify their raw material suppliers and identify to whom they supply products. They must have systems to provide those data to the competent authorities, label or identify products to ensure traceability, and withdraw and recall unsafe food from the market.

The fish industry trades globally in a vast range of species and products and is diverse in comparison to other protein sources. There are hundreds of different species of fish captured with different methods of catching, handling and food safety requirements. A wide range of live, chilled, frozen and value added fishery products are produced and traded within the various distribution chains, which also have their specialised food handling and food safety requirements.

There is a huge and complex international trade in the raw materials and in primary and secondary processed products. Although there are major structural differences between the chains for different types of fish, products and countries, there is also a degree of commonality in information requirements. These information requirements can be categorised as:

— Fundamental to traceability. Each food business has to collect and keep information and make it available to the competent authorities and to other food business operators for the purpose of product withdrawal or recall.

— Specifically required information on the nature of the food and the operations involved. This is information required by law for particular purposes and must be made available to the appropriate authorities.

— Commercially desirable information on the nature of the food and the operations. This information can be requested by food business for different reasons such as ethical, environmental, GMP, quality assurance records, raw material or product standards and specifications, etc.

At present, a lot of information is being recorded in the product information system of the processing and distribution chains. Some of that information passes from one link in the chain to the next, either on the label or in the documents, the rest is held by the producer or the distributor. To properly implement traceability it is necessary to define the physical unit and batch.

A physical unit can be an individual large fish, a box or a tub of fish, a package, case or pallet of products, a freight container. The batch can be a catching day, production date, a shipment, etc. The information attached to the unit is to be the key to traceability, i.e. the product number (code), production date and producer's number makes it possible to trace the product to the producer and look up the required information.

A product can be traced either backward or forward. Backward leads to the origin and history, everything that went to a batch and depends on all links mapping identification (ID) of output batches to ID of input batches. Forward trace explains what happened to a certain batch, all the processes and output batches that the batch in question went into. Keeping track of batches and their properties is the key to implementing chain traceability.

A step forward in the implementation of traceability will be the development of a standard for electronic transmission of data and information. Because of increasing information demands from buyers and consumers, it is no longer practical to transmit all the data physically along with the product. A more sensible approach is to mark each package with a unique identifier, and then transmit or extract all the relevant information electronically. But so far it is common to use other ways like telephone, e-mail and fax for communication between the links.

Definitions of Quality

Definitions of quality as applied to food products vary according to the author. Different qualities with respect to seafood include safety, nutritional quality, availability, convenience and integrity, and freshness quality. The most important is seafood safety. Seafood quality is usually influenced by freshness or degree of spoilage of the raw material or the product. Freshness is considered as one of the most important factors determining quality of fish.

Handling, processing and storage techniques can also affect the quality of fish and fishery products as they can result in the occurrence of defects such as bruises, bloodstains, trimming imperfections, etc. Many studies have been done on spoilage processes of fish stored in ice and frozen storage, considering a whole range of influencing factors, which can affect the fish quality. Among others, these factors are related to fish species, size and seasonal condition, fishing method, handling, processing technology, time and temperature development.

Freshness Parameters

Fish is a perishable commodity. Upon death a series of natural changes start, leading to spoilage affecting the shelf life of the fish. Freshness or the extent of spoilage during storage under chilled conditions is the key determinant of the quality of fish and fishery products. Shelf life of fish is defined as the length of time it is fit for human consumption. Spoilage due to microbial activity is the main limitation of the shelf life of iced or refrigerated fish. Off-odours and off-flavours, slime formation, gas production, discoloration and changes in texture are obvious signs of spoilage.

The development of these spoilage conditions in fish and fishery products is due to a combination of chemical, autolytic and microbiological changes, but the spoilage rate can be reduced by taking preventive measures like icing or keeping a low temperature during storage. According to Bonnell, controlling the temperature of fish is perhaps the most important element in the preservation of fresh fish. The proper cooling of fish has a number of advantages. Firstly, bacterial activity depends very much on temperature, the closer it is to 0°C, the slower the rate of bacterial spoilage. Likewise, enzyme activity also decreases as temperature falls, so the rate of autolytic spoilage is significantly slowed.

Chemical spoilage or development of rancidity can be prevented by rapid handling onboard and storage of products under anaerobic conditions. A rise in product temperature accelerates deterioration and reduces quality. If the rate of deterioration is known, it should be possible to determine the quality at any time by continuously monitoring the time and temperature history of fish post mortem. A number of authors have attempted to derive simple mathematical relationships that provide an acceptable measure of deterioration in fish with a known time/temperature history. These relationships have been used in conjunction with time/temperature recording devices to monitor the deterioration of quality in batches of fish. Post mortem changes in electrical properties of fish skin and flesh are also used in determining the potential shelf life of whole fish, but it is not useful for frozen products.

Evaluation of Defects

Sensory inspection of processed fish is used in fish industry to find defects that have occurred during handling and processing. These defects are well described in the technical specifications for the products. Defects can be related to the condition of the fish flesh, appearance, which includes colour defects (bruises, bloodspots) and dehydration, workmanship defects such as improper packaging and cutting and trimming imperfections, scales, bones, foreign matters, skin and black membrane and the size of fillets. Evaluation of defects is widely used in control of processes and to grade fish for selling or buying purposes.

Methods to Evaluate Fish Freshness

Huss and Bonnell discuss the methods applied to evaluate the freshness of fish, which are divided in two categories, sensory and instrumental techniques. Instrumental methods include biochemical and chemical, microbiological and physical. Each of these methods measure different spoilage indicators in fish and fishery products. Only through a combination of instrumental and sensory analysis can optimal information on the product be obtained.

Microbiological Methods

The activity of microorganisms is the main factor limiting the shelf life of fresh fish. The aim of microbiological examinations is to evaluate the possible presence of bacteria or organisms of public health significance and to give an impression of the hygienic quality of the fish. This includes temperature abuse and hygiene during handling and processing. An estimation of the total viable count (TVC) is used as an index in standards, guidelines and specifications. Specific spoilage organisms (SSO) capable of producing hydrogen sulphide or reducing trimethylamine oxide (TMAO) are considered more useful to estimate spoilage and the remaining shelf life of fish and fishery products.

Biochemical and Chemical Methods

Classical chemical methods for the analysis of total volatile bases (TVB) and trimethylamine (TMA) are used for the determination of fish freshness. TVB only reflect later stages of spoilage, so it is not reliable for the first days of chilled storage of fish. Adenosine triphosphate (ATP) is another indicator of fish freshness. The extent of ATP degradation is expressed as the K value, which is defined as the ratio of the sum of inosine and hypoxanthine concentrations to the total concentration of ATP metabolites. There are other methods related to lipid oxidation such as determination of peroxide, thiobarbituric acid and iodine values.

Physical Methods

Ólafsdóttir et al. describe physical methods based on changes in the electric properties of the fish muscle. Different devices are available to measure electrical properties: the Torrymeter, the Fish tester and the RT-Freshness Grader, but they can only be used for fresh fish. Changes of structure and colour also occur in the fish flesh. Texturometers are used to measure the structural changes. Time-temperature indicators (TTI) are based on using some biological, chemical or physical processes that depend on time and temperature and can give information about the time-temperature history of the food. All methods mentioned above provide information on parameters related to fish freshness, but none of them alone is capable to determine whether a fish is fresh or not.

Sensory Evaluation

Sensory evaluation is the most important method for assessing freshness and quality of fish and fish products. Sensory methods offer a rapid and accurate measurement of perceived attributes providing information about food. Sensory evaluation is defined as the scientific discipline used to evoke, measure, analyse and interpret characteristics of food as perceived by the senses of sight, smell, taste, touch and hearing. Sensory tests can be divided into three groups: discriminative, descriptive and affective. The first two are analytical tests in which a trained panel is used, while the third are subjective consumer tests based on a measure of preference or acceptance.

The most commonly used descriptive tests are structured scaling for quality assessment and profiling for a detailed description of one or more attributes. Sensory evaluation is currently the most important method used for freshness evaluation in the fish sector. The most common sensory method used in Europe is the EU scheme, in which three grades of freshness are established: E, A and B, corresponding to various stages of spoilage. Extra is the highest quality, while B is the level where fish is considered unfit for human consumption. This scheme does not take into account differences between species because only general parameters are used.

The Torry scheme, a ten score system for the evaluation of cooked fish, is another method used for sensory evaluation of fish freshness. The Quality Index Method (QIM) is based on characteristic changes in raw fish that occur in the appearance of eyes, skin and gills, and odour and texture, and a score system from 0 to 3 demerit points.

The scores for all the attributes are summarised to give an overall sensory score, the Quality Index, which is then compared to a QIM calibration curve to establish the relative freshness in terms of storage (predicted) days in ice and to predict the remaining storage life (shelf life). The description of each score for each parameter is listed in the QIM scheme.

The QIM and the schemes have been developed for several fish species e.g. cod, haddock, redfish, herring, saithe, and shrimp. In this study the QIM method is used in the evaluation of redfish.

Field Study at ÚA

The Utgerdarfelag Akureyrar (ÚA) Seafood Group in Akureyri was visited for a two weeks period, from November 18th to the 30th. This company, and in particular the parent company, was chosen for the study, knowing the fact that ÚA Seafood Group is one of the leading companies in Iceland, operating fishing vessels and processing plants. The company's role is catching, processing and selling a variety of high quality, safe seafood products. During the field study at ÚA the documentation of the quality system was accessed and carefully studied.

The production and quality managers explained in situ the processing flow for the different products made from redfish and cod as well, even though cod was not a part of this study. The processing line was visited daily for a better understanding of the process and the quality control. It was studied how the information about the process and products is generated, how this information is transmitted along the chain and how it is recorded and stored for the purposes of traceability. This means that information is available to attend customer claims, recall of products and also to verify the quality and handling of products. Data and information were collected from evaluations conducted and from the records stored in the company's system.

Structure of the Company and Main Products

The group consists of the parent company and four subsidiaries: Laugafiskur ltd., specialising in drying fish; Jökull ltd., concentrating on the processing of wholefrozen pelagic fish; Hólmadrangur ltd., for shrimp processing; and GPG Saltfish ltd., specialising in the production of splitted salted fish. The parent company operates trawlers and processes demersal fish, either on shore or on board factory trawlers. The main species harvested are cod (Gadus morhua), redfish (S. marinus/mentella), Greenland halibut, and prawns (Pandalus borealis).

The main products of the group are frozen seafood and dried fish products. The total production of the company in 2000 was more than 17,000 tons with a value of ISK 5.3 billion. Table 1 shows the production by species/product. ÚA operates a sales and marketing department which is responsible for the group sales of fresh and frozen products. The access to fresh fish allows the company to offer high quality products, supported by a quality assurance system based on HACCP, which ensures product safety. The biggest markets for the group are in the USA and UK. Other important markets are Germany, Nigeria, Japan and Taiwan, France and Denmark.

Table 1: Volume of production of ÚA Seafood Group in 2000.

Product/Species	tons
Cod	6,086
Redfish	2,833
Haddock	287
Saithe	177
Greenland halibut	1,232
Capelin roe	248
Shrimp in the shell	3,224
Cooked and peeled shrimp	1,255
Total frozen products	15,342
Stockfish	2,215
Total	17,557

The ÚA parent company operates five fishing vessels and two land-based processing plants. Three of the vessels are fresh fish trawlers, supplying the factories in Akureyri and Grenivík, one vessel is a freezing trawler for demersal fish, and the fifth vessel is a shrimp-freezing trawler. The fresh fish trawlers concentrate on the catching of cod (G. morhua), haddock (Melanogammus aeglefinus), redfish (S. marinus/

mentella) and saithe (Pollachius virens), while the freezing trawlers focus on Greenland halibut (R. hippoglossoides), redfish (S. marinus/mentella) and shrimp (P. borealis). The ÚA plant in Akureyri is one of the largest of its kind in the North Atlantic and specialises in the processing of cod and redfish. The plant in Grenivík is smaller and processes haddock, saithe and cod.

Quality System

The ÚA Seafood company quality system was studied, especially for redfish. Data and information from the processing line was gathered, including information on the traceability system. For effective quality and process control the plant has implemented a quality system based on HACCP.

Redfish Processing

The redfish (Sebastes marinus/mentella) is a temperate marine species, inhabiting the North Atlantic ocean waters along the European and American coasts. The two species of redfish: ocean perch (S. marinus) and deepwater redfish (S. mentella) are difficult to distinguish. Ocean perch (S. marinus) has a bright red skin and a chubby shape, it can attain a maximum length of 100 cm and weight up to 15 kg, while deepwater redfish attains a maximum size of 55 cm. Redfish is utilised fresh and frozen. It is a good table fish with a firm and tasty flesh. As a result of different storage studies it has been estimated that redfish kept in ice at 0°C has a shelf life of 18 days, granted that the fish was handled under good manufacturing practices on board the vessels.

HACCP System of the Company

The HACCP system, also called own check system is fully described in the HACCP handbook. Control points are established to monitor both product and process in the own check system of the plant. These points are located in the reception area, after trimming, freezing, in casing/labelling/palletising; after battering and breading for breaded products and finally in weighing/packing operation in IQF products.

At each control point samples are taken and evaluated according to the specifications and procedures. Temperature is monitored in every step of the process including frozen storage of final products. All the information generated is recorded electronically or in the corresponding form for each operation or product.

Labelling and Traceability

From the moment the fish is brought on board the vessel information is generated and recorded until the product reaches the end customer. Part of the information will be transmitted from one link to the next, attached to the unit. The rest will remain in the link for internal use and will be accessible if necessary. Most of the information originates in the catching and processing links. Redfish fillets blocks were chosen in our study as an example to illustrate the information recorded during each step of the processing and distribution chain. First it is necessary to identify batches and units within the chain. ÚA receives the raw material to be processed in the plant from its own fishing vessels.

The information about the catch and finished products sent for export is given to the Directorate of Fisheries. Information is transmitted between ÚA and the export company about the fish products available for export. At the request of ÚA the transport company freights the product to frozen storage at the carrier (EIMSKIP). The export company interacts with the shipping company and the buyer in Europe regarding shipments. The export company also provides the information requested by customs about shipment of products. Part of the information is printed out on a label attached to the unit and transmitted to the next link in the production and distribution chain. This information varies according to the product type and packaging.

Quality Evaluation of Fish

Quality evaluation of fish is one of the most important aspects to be considered in the quality system of a fish processing plant. The field study carried out at ÚA provided a good opportunity to perform evaluations, using the information available in the plant and taking samples to conduct the analysis, thus gaining skills and experience.

Evaluation of Raw Materials and Methods

Evaluation of raw material and fillets were carried out based on the quality system of the company. Moreover, sensory evaluation of the raw material using the QIM was done on samples from one fishing trip of the vessel Arbakur.

At ÚA every batch of fish is evaluated at the moment of landing and reception in the plant. A batch in this case is the fish of the same species from all the catch of one fishing trip. The samples are taken randomly from different catching days. This form includes the vessel's name, the number of the fishing trip and date of evaluation. Once the evaluation has been carried out, a new form is filled with the final scores given to the parameters checked. These results are valid for the whole fishing trip (batch). A five score scheme is used. Fish with score 2 or less is unsuitable for processing.

Information about the redfish received from the vessels for a period one month, from September 23rd to October 24th, was accessed and analysed. Whole ungutted redfish from 8 batches was evaluated looking at the handling on board: weight of fish and ice, washing, how the fish was aligned in the tub and icing, i.e. fish-ice layers and ice/fish ratio. The data was recorded and analysed using the principal component analysis technique (PCA) in Unscrambler (Camo, Norway).

After trimming, a random sample of fish fillets are checked for defects in appearance (bones, parasites, bloodspots, bruises, black membrane). Colour and smell are evaluated on a scale from 2 to 5. In this study data collection and evaluation were carried out for redfish fillets after trimming. Data from the quality system on evaluation of fish fillets after trimming were studied. Information was collected from ten processing days, starting from September 26th to October 18th. The raw material came from eight fishing trips of the three vessels that the company owns.

Sensory evaluation of raw material using the QIM method was conducted under processing conditions as part of this study. A batch consisting of three catching days from one vessel's fishing trip was sampled, taking five samples from each catching day. The fishing trip in this case was 6 days long. The samples came from the second, fourth and sixth catching days, it means that they had been in ice for six, four and two days, respectively. The scores for each parameter were determined by consensus of the two untrained assessors evaluating fish, the author of this report and the inspector in the reception area in the plant.

References

Agrawal V.P., *Aquaculture and Fisheries Science*, Delhi, S.R. Scientific, 2006, 630 p.

Datta Munshi J.S. and Singh H.R (ed.), *Advances in Fish Research*, Vol. IV, Delhi, Narendera Pub., 2006, xvi, 312 p.

Khanna D.R. and Yadav P.R., *Biology of Fishes*, New Delhi, Discovery, x, 386 p., 2004.

Saini V.P., Sharma L.L. and Ujjania N.C., *Dictionary of Aquatic Resources and Fisheries*, Udaipur, Agrotech, 2007,. 648 p.

Shafi S.M., *Applied Fishery Science*, New Delhi, Atlantic Pub., 2003, 2 volumes, 601 p..

Singh H.R. and Lakra W.S., (ed.)., *Coldwater Aquaculture and Fisheries*, Delhi, Narendra, 2000, 337 p.

11

Microbial Safety in Food Preservation

Food preservation in the broad sense of the term refers to all measures taken against any spoilage of food. In its narrower sense, however, food preservation connotes processes directed against food spoilage due to microbial or biochemical action. Preservation technologies are based mainly on the inactivation of microorganisms or on the delay or prevention of microbial growth. Consequently they must operate through those factors that most effectively influence the survival and growth of microorganisms.

Factors used for food preservation are called 'hurdles' and there are numerous hurdles that have been applied for food preservation. Potential hurdles for use in the preservation of foods can be divided into physical, physicochemical, microbially derived and miscellaneous hurdles. Among these hurdles, the most important have been used for centuries and are as either 'process' or 'additive' hurdles including high temperature, low temperature, water activity, acidity, redox potential (Eh), competitive microorganisms and preservatives.

Recently the underlying principles of these traditional methods have been defined and effective limits of these factors for microbial growth, survival, and death have been established. Recently, about 50 additional hurdles have been used in food preservation. These hurdles include: ultrahigh pressure, mano-thermo-sonication, photodynamic inactivation, modified atmosphere packaging of both nonrespiring and respiring products, edible coatings, ethanol, Maillard reaction products and bacteriocins. Hurdles used for food preservation influence the quality as well as the safety of foods. The effects of hurdles on food quality can be positive or negative, depending on their intensity.

Some hurdles have antimicrobial properties and at the same time improve the flavor of the products; this also applies to nitrite used in the curing of meat. However, some hurdles also provide a negative effect on food quality. For instance, chilling to an unsuitably low temperature will be detrimental to some foods of plant origin ('chilling injury'), whereas moderate chilling is beneficial for shelf-life. Another example is the pH of fermented sausages, which should be low enough to inhibit pathogenic bacteria but now so low as to impair taste. If the level of a particular hurdle in a food is too small, it should be strengthened; if it is detrimental to the food quality, it should be lowered and an additional hurdle considered.

To ensure the stability and safety of foods, each hurdle should be applied in the proper manner. However, the high strength or concentration of any single hurdle could adversely affect food quality such as loss of nutrients, texture, color, etc. Therefore, if manufacturers are to achieve a product that accommodates consumer preferences while remaining safe and stable, they must achieve synergies through the combined use of these or other hurdles which, while they may not prevent the growth of pathogenic or spoilage microorganisms on their own, will do so collectively. And each product may require a different combination of hurdles, depending on a range of factors including the following;

(i) the initial microbial load of the product requiring preservation,

(ii) how favourable conditions are within the product for microbial growth,

(iii) target shelf-life.

An important current trend is toward the use of procedures that deliver food products are less "heavily" preserved, higher in quality, perceived as being more "natural", contain less additives, and are nutritionally healthier. Some new and "emerging" techniques aim to meet some of these objectives. Most of these techniques act by inactivating microbes.

Many naturally occurring antimicrobials have been explored for used as food additive preservatives. But few have yet been widely exploited. Lysozyme has gained a useful market for the destruction of outgrowing cells from spores of Clostridium tyrobutyricum in some cheeses. The bacteriocin, nisin, likewise has a well-developed portfolio of uses in cheese, canned and some other foods, and the antimycotic natamycin (pimaricin), is employed to prevent mold growth on cheese and surfaces of salami type products.

Hurdle Technology

The combined use of several preservation methods, possibly physical and chemical, or a combination of different preservatives is an age-old practice. It has been commonly applied by the food industry to ensure food safety and stability. In smoked products, for example, combination treatment includes heat, reduced moisture content and antimicrobial chemicals deposited from the smoke onto the surface of the food.

Some smoked foods may also be dipped or soaked in brine or rubbed with salt before smoking, to impregnate the flesh with salt and thus add a further preservative mechanism. In jam and other fruit preserves, the combined factors are heat, a high solids content (reduced water activity) and high acidity. In vegetable fermentation, the desired product quality and microbial stability are achieved by a combination of factors such as salt, acidification, and so forth.

In recent years, the concept of combining several factors has been developed by Leistner and others into the 'hurdle effect'. From an understanding of the hurdle effect, hurdle technology has been derived, which has the goal not just to understand why a certain food is safe and stable, but to improve the microbial quality of the food by an optimisation and intelligent modification of the hurdles present. It employs the intelligent combination of different hurdles or preservation techniques to achieve multi-target, mild but reliable preservation effects.

Hurdle technology has arisen in response to a number of developments;

(i) Consumer demands for healthier foods that retain their original nutritional properties,

(ii) The shift to ready-to-eat and convenience foods which require little further processing by consumers,

(iii) Consumer preference for more 'natural' food which require less processing and fewer chemical preservatives.

Hurdle technology provides a framework for combining a number of milder preservation techniques to achieve an enhanced level of product safety and stability.

Mechanism

Microorganisms react homeostatically to stress factors. When their environment is disturbed by a stress factor, they usually react in ways that maintain some key element of their physiology constant. Microorganisms undergo many important homeostatic reactions. Preservative factors functioning as hurdles can disturb one or more of the homeostasis mechanisms, thereby preventing microorganisms from multiplying and causing them to remain inactive or even die. Therefore, food preservation is achieved by disturbing the homeostasis of microorganisms.

The best way to do this is to deliberately disturb several homeostasis mechanisms simultaneously thus a combination of multiple hurdles (hurdle technology) could increase the effectiveness of food preservation. The success of hurdle technology depends on ensuring metabolic exhaustion. Most stress reactions of microorganisms are active processes, and this often involves the expenditure of energy, e.g. to transport protons across the cell membrane, to maintain high cytoplasmic concentrations of 'osmoregulatory' or 'compatible' solutes.

Restriction of the availability of energy is then a sensible target to pursue. This probably forms the basis of many the successful, empirically derived, mild combination preservation procedures exemplified by hurdle technology. As an example, if a food can be preserved by lowering the pH, then it is sensible also to include a weak acid preservative which will amplify the effect of the protons or to allow a milder, higher pH to be employed. It is sensible if proton export is made more difficult by the additional requirement that cells be forced to regulate osmotic strength. Then, if the food can be enclosed in oxygen-free vacuum or modified atmosphere packaging, facultative anaerobes will be further energy-restricted at a time when the various stress and homeostatic reactions are demanding more energy if growth is to proceed.

However, environmental stresses can provide varying results because some bacteria may become more resistant or even more virulent under stresses through stress reactions such as synthesis of protective stress shock proteins. It has been reported that synthesis of protective stress shock proteins is induced by several stresses including heat, pH, aw, ethanol, oxidative compounds, and starvation. And, although each stress has a different spectrum of antimicrobial action, those stress reactions might have a non-specific effect, since due to a particular stress, microorganisms become also more tolerant to other stresses i.e. 'cross-tolerance'.

For instance, acid-shock or acid-adapted cells became tolerant to a range of other environmental stresses in several pathogenic bacteria including E. coli O157:H7, S. typhimurium, and L. monocytogenes. Conversely, the heat shock response that follows mild heating can result in cells becoming more acid tolerance. Therefore, the various stress responses of microorganisms might hamper food preservation and could turn out to be problematic for the application of hurdle technology when hurdles are used consequitively. However, the use of different stresses at the same time may also prevent the synthesis of those protective proteins because simultaneous exposure to different stresses will require energy-consuming synthesis of several or at least much more protective stress shock proteins which in turn may cause the microorganisms to become metabolically exhausted.

This antimicrobial action of combining hurdles is known as 'multitarget preservation' introduced by Leistner. The concept of multitarget preservation increases the effectiveness of food preservation by using a combination of different hurdles which have different spectra of antimicrobial actions. It has been suspected for some time that combining different hurdles for good preservation might not have just an additive effect on microbial stability, but they could act synergistically.

A synergistic effect could be achieved if the hurdles in a food hit, at the same time, different targets within the microbial cells and thus disturb the homeostasis of the microorganisms present in several respects. Thus repair of homeostasis as well as the activation of stress shock proteins becomes more difficult. Therefore, simultaneously employing different hurdles in the preservation of a particular food should lead to optimal microbial stability.

In addition, no one preservative factor is active against all the spoilage microorganisms present in foods. An attempt is therefore made to compensate for this deficiency by combining various preservative factors having different spectra of action. Since from this multitargeted approach, hurdle technology could more effective than single targeting, it allows the use of individual hurdles of lower intensity for improving product quality as well as for food preservation.

Limitations of Hurdle Technology

As described above, hurdles used in food preservation could provide varying results depending on bacterial stress reactions such as the synthesis of protective proteins. These stress reactions or cross-tolerance may not exist when combined hurdles are used. However, although hurdles are applied simultaneously in combined form, there are three possible results whereby the action may be changed by combining two or more preservative factors:

(i) addition or additive effect,

(ii) synergism or synergistic effect,

(iii) antagonism or antagonistic effect.

The term additive effect denotes that the effects of the individual substances are simply added together. Synergistic effect is the expression used when the inhibitory action of the combination is reached at a concentration lower than that of the constituent substances separately. An antagonistic effect is the opposite of this latter, i.e. one where the mixture concentration required is higher than that of the individual constituents. Among these results, first two are desirable results and the main reason the hurdle technology is employed for food preservation rather than one hurdle.

Generally, it is accepted that the combination of hurdles has a higher inhibitory effect than any single hurdle. However, recently, some studies showed that combination treatments were less effective at reducing levels of microorganism than were single treatment. These effects of combining hurdles were antagonistic. In some cases, application of the hurdle concept for food preservation may inhibit outgrowth but induce prolonged survival of microorganisms in foods.

The various responses of microorganisms under mild stress conditions of hurdle technology might hamper food preservation and could turn out to be problematic for the application of hurdle technology. However, no general statements can be made about the actions of any particular preservative method on other factors. Additionally, in many studies, any synergistic effects observed in laboratory tests were so weak as to have no significance for practical food preservation. It was mainly for commercial reasons that most combination products formerly marketed in large numbers were preferred to straight preservations. Although some combinations of hurdles showed less or not significant effectiveness for killing microorganisms in foods, many promising hurdles have been identified so far. However, the application of the idea in the food industry has been largely restricted to the meat sector.

Recent studies, however, emphasize a much wider potential application, e.g., in bakery products, fish, and dairy products, More specifically, the concept was introduced into mild processing of fruits and vegetables. However, there is only limited information available about the effect of combined hurdles in these types of foods. The vast majority of preserved foods that are consumed in different countries rely on combinations of preservative factors for their stability and microbiological safety. Therefore, it will be important to understanding the interactions of different hurdles in various foods to find types and intensity of that hurdles are needed for the desired microbial safety and stability of a particular food.

Pickling Methods

Pickling is an age-old method for preserving vegetables and fruits. The manufacture of pickles, relishes, and condiments has become one of the most important food industries. Although the preservation of vegetables and fruits in pickled form began as a household art, at present most of the world's supply of pickles is produced in commercial plants. Any vegetables or fruit may be pickled, although the quality of some pickled foods is poor. The cucumber, one of the most important raw material used for pickling, is packed in many forms, e.g., in plain or spiced sweet vinegar in jars, kegs, or cans or fermented in spiced brine as dill pickles; packed in mustard; or in chopped form in various relishes.

Green tomatoes, peppers, cauliflower, and onions are common ingredients of mixed pickles, chowchow, etc. Moreover, sauerkraut and olives are also important pickled products. Recently, more types of fruits and vegetables have been used to produce pickled products, including asparagus, beets, peaches, figs, pears, and so on. Ingredients used for producing pickled products include salt, vinegar (acetic acid), spices, sugar, and water. Pickled fruits and vegetables are made by immersing raw materials in brine containing vinegar (acetic acid) and salt, and then heat treated.

Spices and sugar are used as additional ingredients to improve flavor. Therefore, salt, acetic acid, and heat are considered as major factors for increasing the microbial safety of pickled products. However, sometimes other preservatives are also added to enhance microbial safety. However, the U.S. Food and Drug Administration regulations do not allow the use of preservatives as the primary barriers to the growth of microbial pathogens in acidified foods.

For these food products, the Code of Federal Regulations states only that acid or acid ingredients which will be acetic acid must be added so that the pH is maintained at or below 4.6; a heat treatment must be included in the process, if necessary, to prevent the growth of microbial pathogens. These regulations were designed to control the growth and toxin production by Clostridium botulinum. The regulations do not take into account the amount or type of organic acid present in acidified foods.

For thermal processing, the pickle industry usually uses a procedure established in the early 1940s that recommended an internal pasteurizing temperature of 74°C for 15 min followed by prompt cooling. Most companies, however, have developed their own processes for specific products. Heat processing or pasteurisation in pickling is performed to increase microbial food safety. However, some vegetables used in pickling are very heat sensitive and quality indicators such as color and texture are usually degraded to a large extent during thermal treatment. Therefore, some pickled vegetables such as freshpack pickles are produced without heat processing.

Moreover, consumers insist upon the freshness of food products. The demand for high-grade, mild (low salt and low acid), refrigerated pickles has increased over the past 10 years, and the pickling industry has responded with a range of products. Overnight dills or refrigerated dills might be an example of pickled products produced without heating. The refrigerated dill is essentially a non-heated, well-acidified, low-salt content, refrigerated green cucumber, containing one or more preservatives with spices and flavouring. Typically, overnight dills are made from the freshest cucumbers which are washed and packed by hand into containers.

The cucumbers are then covered with a brine consisting of water, acetic acid (vinegar), and salt. The equilibration pH desired is 4.2 to 4.3 with a titratable acidity of 0.3 to 0.5% as acetic acid. Always dill, and usually garlic, along with other flavouring agents, plus sodium benzoate as a preservative are added. The containers are closed, cased, and moved to a refrigerated warehouse. Overnight dills are kept refrigerated in marketing channels until purchased by the consumer.

References

Arnold, K. W., and C. W. Kaspar., "Starvation- and stationary-phaseinduced acid tolerance in *Escherichia coli* O157:H7," *Appl. Environ. Microbiol.*, 1995, 61:2037-2039.

Beuchat, L. R., "Synergistic effects of potassium sorbate and sodium benzoate on thermal inactivation of yeasts," *J. Food Sci.*, 1981a, 46:771-777.

Desrosier, N. W., *The Technology of Food Preservation*, The AVI Publishing Company, Inc. Westport, Connecticut, 1959, pp.52-98.

Leistner, L., "Basic aspects of food preservation by hurdle technology," *Int. J. Food Microbiol*, 55:181- 186, 2000.

Matin, A., E. A. Auger, P. H. Blum, and J. E. Schultz., "Genetic basis of starvation survival in nondifferentiating bacteria," *Annu. Rev. Microbiol.*, 1989, 43:293-316.

12

Food Safety Risk Analysis

The importance of risk assessment lies not only in its ability to estimate human risk, but also in its use as a framework for organising data as well as for allocating responsibility for analysis. Within common boundaries, it is important to understand that risk assessment is a process that can include a variety of models to reach conclusions. The concept of the Acceptable Daily Intake (ADI), for example, may be considered as a component of risk assessment, although a notionally zero risk, when combined with exposure assessments.

Common structure permits a transparent and relatively uniform approach to the problem of providing useful information to risk managers. Moreover, for the evaluators, it provides a way to identify areas in which the available data are insufficient to reach reasonable decisions. The use of a common framework also facilitates harmonisation of food safety evaluation processes among nations.

Risk assessment is seen primarily as a method of systematically organising scientific and technical information, and its associated uncertainties, to answer specific questions about health risks. It requires evaluation of relevant information, and selection of the models to be used in drawing inferences from that information. Further, it requires explicit recognition of uncertainties and, when appropriate, acknowledgement that alternative interpretations of the available data may be scientifically plausible. Risk assessment is subject to uncertainties related to data and to the selection of the appropriate model.

The process of risk assessment requires adequate toxicological information preferably based on standardised testing protocol which have been accepted by the international community. Depending upon the chemical, empirically-based answers to toxicological questions may be available for the purpose of risk assessment. However, in no case will the scientific information be comprehensive enough to provide a high degree of certainty. When several sets of animal toxicology data are available, there are usually insufficient data to identify the set that best predicts human response. As a result, it has become traditional to rely on toxic responses which occur at the lowest dose in a study of acceptable quality.

Minimum data requirements for risk assessment are difficult to specify in advance. Hazard, dose-response, and exposure data bases for substances that may become subjects for risk assessment vary enormously in size, scope, and quality. In some instances, the data may be very limited and practically impossible to obtain. The latter is especially the case for contaminants and naturally occurring substances. When a risk assessment is necessary, risk assessors are required to make the best use of whatever information is available, and to deal explicitly with data uncertainties. In cases where this is not possible, risk assessors should provide the reasons for ·such judgements. Perhaps the appropriate option is to leave the question of minimum data requirements open to such case-by-case judgements.

Other issues related to the process of risk assessment include the use of default assumptions to fill knowledge and data gaps. This provides the advantage of ensuring consistency in approach and minimising or eliminating case-by-case manipulations of the conduct of risk assessment to meet predetermined risk management objectives. One major disadvantage, however, is the potential for displacement of scientific judgement by rigid guidelines. One intermediate approach is to allow risk assessors to replace defaults in specific cases of chemicals for which relevant scientific data are available to support alternatives. Specific and explicit justification for any such departures should be provided.

Effect of Hazard Identification

The goal of hazard identification is to identify potential adverse health effects in humans associated with exposure to a chemical, the likelihood of such effects occurring and the certainty or uncertainty associated with such effects. The hazard identification does not imply the quantitative extrapolation of risk for exposed human population as in the dose-response and risk characterisation step, but rather an evaluation of the qualitative likelihood of the effect occurring in exposed human population. Because data are often insufficient, hazard identification is best conducted using the weight-of-evidence approach.

The approach requires an adequate and documented review of relevant scientific information obtained from appropriate databases, peer-reviewed literature and, if available, unpublished studies from other sources, such as industry. This approach places emphasis on studies in the following order: epidemiological studies, animal toxicological studies, in vitro assays and, lastly, quantitative structure-activity relationships.

Epidemiological Studies

Where data from positive epidemiological studies are available, their use in the risk assessment process is encouraged. Data derived from human clinical studies, where they are available, should also be utilised in the hazard identification step, as well as perhaps other steps. However, clinical and epidemiological data are unlikely to be available for most chemicals.

In addition, negative epidemiological data may be difficult to interpret for risk assessment purposes because the statistical power of most epidemiological studies is inadequate to detect effects at relatively low levels in human population. Finally, although the value of epidemiological data is recognised, positive data indicate that an adverse effect has already occurred; thus, risk management decisions should not be delayed pending the development of epidemiological studies. Epidemiological studies from which data for risk assessment are derived should be based on recognised standardised protocol.

During the design of epidemiological studies, or where positive epidemiological data are available, consideration must be given to the variability in human susceptibility; genetic predisposition, age-related and gender-related susceptibility, and the impacts of factors such as socio-economic status, nutritional status, and other possibly confounding factors. Due to the cost of epidemiological studies and due to the paucity of data such studies provide, hazard identification will ordinarily need to rely on data derived from animal and in vitro studies.

Animal Studies

Most toxicological data for risk assessment are derived from animal studies and it is, therefore, essential that these studies be conducted following widely accepted, standardised testing protocol. While many such protocol exist, e.g. OECD, EPA, guidance is not available concerning the selection and use of specific protocol for food safety risk assessment. Regardless of which protocol are used, all studies should follow Good Laboratory Practices (GLP) and standardised quality assurance/quality control (QA/QC) procedures.

Adequate minimum data sets are generally available for food safety risk assessment and should be used. These include specification of the number of species/strains/stocks, use of more than one sex, appropriate selection of doses, route of exposure, and adequate sample size. In general, the source of data (published studies, unpublished studies, corporate data, etc.) is not a point of great concern as long as studies are transparent and can be demonstrated to conform to GLP and QA/QC procedures.

Animal data from long-term (chronic) studies are critical, and should address significant toxicological effects/end-points, including cancer, reproductive/developmental effects, neurotoxic effects, immunotoxic effects, and others. Animal data from short-term (acute) toxicity studies will also be useful and should be generated. Animal studies should facilitate identification of the range of toxicological effects/end-points (including those listed). Data on the relationship between toxicity and essentiality should be gathered for those substances which are required to meet nutritional requirements, e.g. copper, zinc, and iron. Animal toxicological studies should be designed to identify a NOEL, a no-observed-adverse-effect level (NOAEL) or a benchmark dose; that is, doses should be selected to identify these end-points.

Doses should also be selected at levels high enough to reduce the likelihood of false-negatives as much as possible, while considering issues such as metabolic saturation, cytogenic and mitogenic induced cell proliferation, etc. Presently, the selection of the highest dose for chronic rodent bioassays is being debated.

Animal studies should, where possible, identify not only potential adverse effects for human health but also provide information on the relevance of these effects for human risk. Information on relevance may be provided by studies that characterise the mechanism of action, the relationship between administered and delivered dose, and by pharmacokinetic and pharmacodynamic studies. Mechanistic data may be supplemented by data from in vitro studies, such as information on genotoxicity derived from reversion assays or other similar assays. These studies should be conducted following GLP, and other widely accepted protocol. However, data from in vitro studies should not be used as the sole source of information to predict human risk.

The results of in vivo and in vitro studies can enhance the understanding of mechanisms and pharmacokinetics/dynamics. However, such information may not be available in many cases and the risk assessment process should not be delayed pending development of mechanistic and pharmacokinetic/dynamic data.

Information on administered versus delivered dose will be useful as part of the evaluation of mechanism and pharmacokinetic data. The assessment should also consider information on chemical speciation (administered dose) and metabolite toxicity (delivered dose). As part of this consideration, the issue of chemical bioavailability should be addressed (bioavailability of parent compound, metabolites, etc.) with specific consideration given to absorption across the appropriate membrane (i.e., the gut), transport to systemic circulation, and, ultimately, to the target organ. Finally, structure-activity relationships may be useful to increase the weight-of-evidence for human health hazards identification. Where classes of compounds are of interest, and where adequate toxicological data are available on one or more members of the class, a toxic equivalence approach may be useful to predict the human health hazard associated with exposure to other members of the class.

Characterisation of Hazard

The chemicals in food being considered include food additives, pesticides, veterinary drugs and contaminants. They are often present in food at low levels—typically at a part per million or less. However, to obtain adequate sensitivity, animal toxicological studies must be conducted at high levels which may exceed, depending on the intrinsic toxicity of the chemical, several thousand parts per million. The significance that the adverse effects detected in high-dose animal studies have for low-dose human exposures is the major question posed in the hazard characterisation of chemicals.

In order to be compared to human exposure levels, animal data need to be extrapolated to doses much lower than those studied. This extrapolation procedure is uncertain both qualitatively and quantitatively. The nature of the hazard may change with dose or may disappear entirely. The selected dose-response model may be incorrect if the nature of the response in animals and humans is qualitatively the same. Not only is the equivalent dose estimate in animals and humans a problem in comparative pharmacokinetics, but also is the change in metabolism with dose.

The metabolism of chemicals at high and low doses may differ. For example, high doses often overwhelm normal detoxification/metabolism pathways and produce adverse effects that would not occur at lower levels. High doses can induce higher rates of enzyme production, physiological changes and dose related pathological changes. The toxicologist must consider the potential impact of these and other possible dose-related changes on the extrapolation of the adverse effect to lower doses.

Toxicologically equivalent doses in animals and humans is a debatable issue. JECFA and JMPR have typically used mg per kg of body weight for interspecies scaling. Recently regulatory authorities in the USA have proposed a scaling equivalent to mg per 3/4 kg of body weight, which is based on more recent pharmacokinetic information. The ideal scaling factor would be obtained by measuring tissue concentrations and clearance rates in the target organ of the animal and human; blood levels would approximate this ideal. Generic interspecies scaling factors should be recognised as default values that are used in the absence of better information, which is seldom available.

Traditionally, toxicologists have accepted the existence of thresholds for adverse effects with the exception of carcinogenicity. The tradition extends from the early 1940s when it became evident that the initiating event in carcinogenesis could be a somatic mutation. In theory, a few molecules, even a single molecule, could cause a mutation that could persist in the animal or human and ultimately be expressed as a tumour. Theoretically, there may be no safe dose for a carcinogen that acts through this mechanism.

In recent years it has been possible to discriminate between carcinogens and to identify a category of non-genotoxic carcinogens that are themselves not capable of producing mutations but act at later stages of the cancer process on cells already "initiated" by other carcinogens or other processes e.g. radiation. In contrast, other carcinogens induce genetic alterations in somatic cells with activation of oncogenes and/or inactivation of cancer suppressor genes. Thus, genotoxic carcinogens are defined as chemicals which can cause genetic alterations in target cells, either directly or indirectly.

While the major target of genotoxic carcinogens is genetic material, non-genotoxic carcinogens act at extra-genetic sites, leading presumably to enhanced cell proliferation and/or sustained hyperfunction/dysfunction at the target sites. Regarding species differences in carcinogenic effects, a large body of data has been reported indicating that quantitative differences exist in both genotoxic carcinogens and non-genotoxic carcinogens. In addition, certain non-genotoxic carcinogens, called rodent-specific carcinogens can be raised as examples of substances for which there are qualitative differences in the ultimate carcinogenic effects. In contrast, no such clear-cut examples have been reported for genotoxic carcinogens.

Toxicologists and geneticists have devised tests to detect chemicals capable of causing mutations in DNA; the Ames test is a well known example. Several such tests, both in vitro and in vivo tests are used, typically in the form of a battery, to determine the mutagenic potential of chemicals. While the exact tests to include in such a battery may be debatable, in general these tests have been useful in distinguishing between genotoxic and non-genotoxic carcinogens.

Food safety authorities in many countries now make a distinction between genotoxic and non-genotoxic carcinogens. While this distinction cannot be applied in all instances due to insufficient information or knowledge on carcinogenesis, the concept can still contribute to the establishment of evaluation strategies for cancer risks posed by exposure to chemicals. In principle, non-genotoxic carcinogens may be regulated using a threshold approach, such as the "NOEL-safety factor" approach.

A safe level or Acceptable Daily Intake (ADI) is derived from an experimental NOEL or NOAEL by applying appropriate safety factors. The conceptual basis for their use is that thresholds will exist at reasonably comparable doses in both humans and experimental animals. For humans, however, sensitivity may be greater, genetic outbreeding may be larger and dietary habits may be more variable. As a consequence, a safety factor is applied by JECFA and JMPR to take into account these uncertainties.

A safety factor of 100 is typically applied when data from long-term animal studies are available but other safety factors are used by different health agencies. JECFA also uses a larger safety factor when the data are minimal or when the ADI is assigned on a temporary basis. Other health agencies adjust the ADI for the severity or irreversibility of the effect. These differences in ADI values constitute an important risk management issue which deserves some attention by appropriate international bodies.

The message communicated with an ADI is that there is no significant risk if the chemical is ingested at or below the ADI. The safety factor, as indicated, is selected to subsume anticipated variations in human responses. It is, of course, theoretically possible that some individuals are even more sensitive than provided for by the safety factor.

Another approach to ADI development has been to move away from reliance on the NOEL/NOAEL and toward the use of a lower effective dose, such as ED 10 or ED05. This approach, called the benchmark dose, draws more heavily on data near the observed dose-response range, but is still subject to the application of safety factors. Thus, while it may allow a more accurate prediction of low dose risk, the benchmark dose-based ADI may not differ significantly from a NOEL/NOAEL-based ADI. Special population groups, like children, are protected by an appropriate choice of the intraspecies conversion factor and by special consideration of their exposures, if necessary.

For genotoxic carcinogens, the "NOEL-safety factor" approach is generally not considered a suitable method for setting acceptable intake levels. The consensus is predicated on the anticipated presence of risk at all doses, even the lowest. At this point, two management approaches are available:

— to ban the chemical from commercial use, or

— to establish a level of risk that is sufficiently small to be deemed negligible, insignificant or societally acceptable.

The implementation of this latter approach has given rise to quantitative risk assessment for carcinogens. Various extrapolation models have been utilised for this purpose. Currently models use experimental measurements of tumour incidence and dose and virtually no other biological information. None of these models have been validated beyond the experimental range. No correction for high dose toxicity, enhanced cellular proliferation, or DNA repair is made. For these reasons, the current linear models are considered to be conservative estimates of risk. This is usually expressed by characterising the risks generated by such models as "plausible upper bounds" or "worst-case estimates".

It is acknowledged by many regulatory agencies that actual or probable human risks are not being predicted. Some countries attempt to reduce the conservatism inherent in linear extrapolation by using non-linear models. An essential component of this approach is the determination of an acceptable risk level. In the USA, FDA and EPA have chosen a risk level of one in a million (10^{-6}). This acceptable level was chosen because it was considered to represent an insignificant risk. But the choice of a risk level is ultimately a risk management decision for each country to decide.

For food additives and residues of pesticides and veterinary drugs, a fixed level of risk is practical as the substances can be disallowed if the estimated risk exceeds the regulatory acceptable level. But for contaminants, including discontinued pesticides which have become environmental contaminants, an established acceptable level can easily be exceeded.

Influence of Exposure Assessment

Estimates of dietary intakes of food additives, residues of pesticides and veterinary drugs and contaminants require information on the consumption of relevant foods and the concentrations of the chemical of interest in those foods. In general, three approaches are available in exposure assessment:

— total diet studies;

— selective studies of individual foods, and;

— duplicate portion studies.

Guidelines for the study of dietary intakes of chemical contaminants are available from WHO. In recent years, direct monitoring of human tissues and body fluids has been increasingly used to assess exposure. For example, the determination of levels of organochlorine compounds in breast milk, which are mainly derived from the diet, has provided an integrated assessment of human exposure to these substances.

Dietary intake determinations can be relatively straightforward for additives, pesticides and veterinary drugs as the relevant foods and their use levels are specified by their approved conditions of use. However, the actual levels of additives and residues of pesticides and veterinary drugs present in foods are often well below the maximum levels permitted. In regard to residues of pesticides and veterinary drugs, levels on or in food are often totally absent because only a portion of the crop/animal population is usually treated. Data on the levels of food additives in foodstuffs can be obtained from the manufacturers. The dietary intake of contaminants requires information on their distribution in foods that can only be obtained by analysing representative samples of foods with sufficiently sensitive and reliable analytical methods. Guidelines for establishing or strengthening national food contamination monitoring programmes have been elaborated.

Maximum Residue Limits (MRLs) for pesticides and veterinary drugs and Maximum Levels for additives can be established from their conditions of use. In the simplest case, a food additive used at a specific level would be stable in the food until consumption. The Maximum Level would then equal the intake level. However, in many cases, the amount of the chemical of interest may change prior to consumption. For example, food additives may degrade during storage or react with the food. Pesticide residues in raw agricultural products may degrade/accumulate during further processing. The fate of veterinary drug residues in food products is influenced by metabolism, kinetics, distribution and withdrawal periods required for treated animals.

The establishment of MRLs must take into account any changes in the nature or level of the residue that may occur prior to a commodity entering commerce or that may occur under any anticipated conditions of subsequent use. Contaminants have no intended technicological effect in the food and guideline levels are usually set as low as reasonably achievable.

The theoretical total dietary intake of additives, pesticides and veterinary drugs must be below their corresponding ADIs. Frequently, the actual intake is well below the ADI. Setting guideline levels for contaminants present special problems. There is usually a paucity of data to establish a provisional tolerable intake. On occasion, the levels of the contaminants are higher than what an established provisional tolerable intake would permit. In these cases, the guideline levels are set on economic and/or technical considerations.

Reliable food intake data are essential for exposure assessments based on measuring levels of chemical agents in food. Detailed food consumption data for the average and median consumer as well as for different population groups are important for assessing exposure, particularly by sensitive groups. In addition, comparable food consumption data, particularly with respect to staple foods from different regions of the world are essential for developing an international risk assessment approach to food safety.

GEMS/Food currently maintains a database of five regional diets as well as a composite "global" diet. Daily dietary intakes of nearly 250 individual primary and semi-processed food commodities are available. The African, Asian, East Mediterranean, European and Latin American regional diets are based on selected national data from FAO Food Balance Sheets. Consumption data derived using this approach provide no information on extreme consumers. No information is available in GEMS/Food on the intake of food additives although intakes in developed countries are anticipated to be greater than in developing countries because of the higher portion of processed foods in the diet.

Risk Characterisation

The outcome of the risk characterisation is an estimate of the likelihood of adverse health effects in human population as a consequence of the exposure. The risk characterisation is performed by taking into consideration the results of the hazard identification, hazard characterisation, and exposure assessment. For threshold acting agents, population risk is characterised by comparison of the ADI (or other measures) with exposure. In this case, the likelihood of adverse health effects is notionally zero when exposure is less than the ADI. For non-threshold acting agents, population risk is the product of exposure and potency.

At the risk characterisation step, the uncertainties involved in each step of the risk assessment process should be described. Uncertainty in risk characterisation will reflect the uncertainties in the preceding steps. The extrapolation of results of animal studies to the human situation may produce two types of uncertainties:

— uncertainties with respect to the relevance of the experimental findings to the humans. For example, forestomach tumours in rats fed butylated hydroxyanisole (BHA) and neurotoxic effects in mice produced by aspartame may not have human parallels; and,

— uncertainties with respect to specific human sensitivity for effects of a chemical that cannot be studied in experimental animals.

In this case, hypersensitivity to glutamate is an example. In practice, these uncertainties are dealt with by expert judgement and by additional studies, preferably in humans. These studies may be performed during the pre-marketing phase as well as during the post-marketing phase.

Biological Risk Assessment

Food will always present some minimal biological risk and it is the task of the food industry to maintain the level of risk at the minimum which is practical and technologically feasible. It should be the role of official bodies to use risk analysis to determine realistic and achievable risk levels for foodborne hazards and to base food safety policies on the practical application of the results of these analyses. Biological agents (hazards) of concern to public health include pathogenic strains of bacteria, viruses, helminths, protozoa, algae, and certain toxic products they may produce.

Of these hazards, the presence of pathogenic bacteria in foods currently presents the most significant problems internationally. The analysis of risk associated with bacterial pathogens presents unique features for risk assessment. There is a need to ensure these hazards are eliminated or reduced to an acceptable level and the CAC and its subsidiary bodies have elaborated a multitude of standards and Codes of Practice which contain procedures designed to address this issue. However, the need for a more formal approach to the process of controlling biological hazards has been recognised by Codex. In adopting the Codex text on the Hazard Analysis Critical Control Point (HACCP) system, the CAC acknowledged that the HACCP system was the most cost-effective method devised to date for controlling foodborne hazards.

HACCP is a system which identifies specific hazards and preventative measures for their control. The seven principles of HACCP, as adopted by Codex, establish the framework for developing specific HACCP plans for each food product/ production line combination. When developing a specific HACCP plan, the identification of all potential hazards which are "of such a nature that their elimination or reduction to acceptable levels is essential to the production of a safe food" is required. However, the determination of which potential hazards are "essential" to control, will involve a risk-based hazard assessment. This hazard assessment will result in a list of the significant hazards which should be addressed within the HACCP plan.

Biological Hazards

Risk assessment is the scientific evaluation of the known or potential adverse health effects resulting from human exposure to foodborne hazards. It provides an estimate of the severity and likelihood of harm resulting from exposure to a hazard. When one is considering public health risks posed by chemical or biological hazards in food, the objective of a quantitative risk assessment is to derive a mathematical statement, based on the probability of certain events, of the chance of adverse health consequences resulting from exposure to an agent capable of causing harm. Possible approaches for conducting a risk assessment for foodborne pathogens include probabalistic scenario analysis, fault tree analysis, event tree analysis and following the quantitative risk assessment paradigm proposed for chemicals.

Biological hazards may act through two general mechanisms in causing human illness. One mode of action is to produce toxins which may cause effects that range from mild symptoms of short duration to severe intoxications that can have long-term or life-threatening consequences. The second mode of action is to produce pathological responses that result from ingestion of viable organisms capable of infecting the host. Threshold levels for concern are easier to quantify in the former case.

In these cases, as with certain other biological agents, a quantitative risk assessment may be possible. When considering hazards from pathogenic bacteria, however, a qualitative risk assessment may be the only feasible method currently available to derive an assessment of the severity and the likelihood of harm associated with exposure through ingestion of a food. However, both the quantitative and qualitative methods will depend on the type and quality of information developed during the risk assessment process.

While the basic steps are the same, their application will be different when conducting a qualitative assessment, as the analyst will not have the information necessary to develop a mathematical estimate of the probability and/or severity of an adverse consequence. When assessing risk for some biological agents, sufficient data may be available to conduct a quantitative analysis. However, the analyst will find that in most cases the many uncertainties associated with how and when an organism may express pathogenic potential will make a quantitative assessment impractical.

In the absence of quantitative data to develop an exposure assessment, measurement of hazard levels at particular process steps, or segments of the production chain, may provide a qualitative indication of likely risks to consumers. In this respect comparative studies on biological hazard levels and qualitative estimates of the likely effects of differences to human health can be used.

Bacterial Hazards

The analysis of risk associated with bacterial pathogens presents unique challenges. Any method used to assess the risk of hazards from foodborne bacteria will be complicated by factors resulting from methods used to grow, process, store and prepare food for consumption. These can vary greatly depending on cultural and geographical differences. Such factors describe the scenario set for a given food and are an essential element in a risk assessment for bacterial hazards. In many cases sufficient data will not be available to support a quantitative assessment of risk associated with pathogenic bacteria. The following present an indication of the type of challenges that make quantitative risk assessment difficult for pathogenic bacteria associated with foods.

Hazard Identification

Bacterial agents known to cause foodborne disease have been identified by using epidemiological and other data to link the organism and its source to illness. However, as only a limited number of outbreaks are adequately investigated, it is likely that a number of bacterial pathogens in food remain to be identified. Limitations on hazard identification include,

— the expense and difficulty involved in outbreak investigations;

— the lack of reliable or complete epidemiological data; and,

— the inability to isolate and characterize new pathogens.

Hazard Characterisation

The purpose of this step is to provide a qualitative or quantitative estimate of the severity and duration of adverse effects due to the presence of a pathogen in food. Dose response data are useful when addressing toxigenic bacteria. However, when characterising hazards from invasive strains of pathogenic bacteria, such information may be of little utility. For many foodborne pathogenic bacteria, dose-response data are limited or non-existent. Information on which to base dose-response estimates is difficult to obtain and may also be inaccurate for a variety of reasons, such as:

— host susceptibility to pathogenic bacteria is highly variable;

— attack rates from a specific pathogen vary widely;

— virulence of a pathogenic species is highly variable;

— pathogenicity is subject to genetic variation resultant from frequent mutation;

— antagonism from other bacteria in foods or the digestive system may influence pathogenicity; and,

— foods will modulate the ability of bacteria to infect and/ or otherwise affect the host

Exposure Assessment

An exposure assessment will give an estimate of either the number of pathogenic bacteria or the level of bacterial toxin consumed in food. While levels of chemical agents in food may change slightly due to processing, population of bacterial pathogens are dynamic and may increase or decrease dramatically in food matrices. Changes in population of bacteria are affected by complex interactions of factors such as those listed below:

— ecology of the bacterial pathogen of concern;

— processing, packaging and storing of food;

— preparation steps, such as cooking, which may inactivate bacterial agents; and,

— cultural factors relating to consumers.

Risk Characterisation

Characterising the risk associated with biological pathogens will depend on the considerations and information described in the hazard identification, hazard characterisation and exposure assessment steps. A risk characterisation will result in a qualitative or quantitative estimate of the potential for adverse effects from a particular bacterial agent on a specific population. It has not yet been determined whether a quantitative risk assessment approach is possible and appropriate for characterisation of risk associated with foodborne bacterial pathogens. Thus, by default, the qualitative approach to characterising risk may be the only current alternative.

The qualitative risk assessment process depends on experience with a specific food, a knowledge of ecology of bacterial pathogens, epidemiological data, and expert judgement regarding hazards associated with the manner in which the food is produced, processed, stored, and prepared for consumption.

Other Biological Hazards

The steps in risk assessment for biological hazards other than bacteria that must be followed, for example, to provide input into HACCP plans. The variations come within the steps where the particular properties of the biological agents sometimes make the assessment more simple or more difficult in terms of developing quantitative data. The factors that complicate risk assessment for foodborne biological hazards (other than bacterial hazards) vary, but are not as great as those for bacterial hazards, as the agents typically do not increase in number when in food.

Role of the CAC in Risk Assessment

Through the CAC and its subsidiary bodies, standards, guidelines and recommendations are elaborated which are designed to address food safety issues related to biological, chemical, and physical hazards. Risk assessment is a key to developing meaningful food safety standards. For the chemical hazards, standards are generally expressed in numerical values based on scientifically derived Acceptable Daily Intake levels. In the case of biological hazards, Codex has elaborated standards, guidelines and recommendations which describe processes and procedures, the application of which supposedly eliminates hazards or reduces them to acceptable levels.

For many of the procedures, there is little quantitative data or other scientific evidence to link specific procedures with potential biological hazards, or with a specific health outcome. These linkages are necessary to enable validation of procedures and processes in standards and Codes that might come before the WTO in adjudicating disputes on food safety measures.

The current procedures and processes contained in Codex standards, guidelines and recommendations do not permit a comparison of relative risk or comparative risk between food safety hazards. There is no established means of comparing chemical hazards with biological hazards, or for comparing biological hazards. This may result in the choice of alternatives which increase overall health risk associated with the food.

Uncertainity and Variability in the Risk Assessment Process

As applied to hazardous agents in food, health-risk assessment is a quantitative evaluation of information on potential health hazards from exposure to various agents and involves four inter-related steps discussed earlier, namely,

— hazard identification;

— hazard characterisation;

— exposure assessment, and;

— risk characterisation.

There are many sources of both uncertainty and variability in the process of human health-risk assessment. While effective risk management policies are possible under conditions of both uncertainty and variability, such policies must take both into account. An uncertainty analysis is an important component of risk characterisation. It provides a quantitative estimate of value ranges for an outcome, such as estimated numbers of health effects. The ranges in the outcome are attributable to the variance and uncertainties in data and the uncertainties in the structure of any models used to define the relationship between exposure and adverse health effects.

Uncertainty and Variability

One of the issues in uncertainty analysis that must be confronted is how to distinguish between the relative contribution of variability and true uncertainty to the characterisation of predicted population risk. Variability refers to quantities that are distributed within a defined population, such as: food consumption rates, exposure duration, and expected lifetime. These are inherently variable and cannot be represented by a single value, so that we can only determine their moments with precision.

In contrast, true uncertainty or model-specification error refers to a parameter that has a single value, which cannot be known with precision due to measurement or estimation error. Variability and true uncertainty may be formally classified as follows:

— Type A uncertainty that is due to stochastic variability with respect to the reference unit of the assessment question, and;

— Type B uncertainty that is due to lack of knowledge about items that are invariant with respect to the reference unit of the assessment question.

There are situations in which true (Type B) uncertainty is negligible relative to variability (Type A uncertainty). In these situations, the outcome of a variance propagation analysis represents the expected statistical variation in dose or risk among the exposed population. When neither variability nor uncertainty are negligible, the shape of the distributional curve representation of variability is unknown because of uncertainties.

Model Uncertainty and Input (parameter) Uncertainty

Uncertainty in model predictions arises from a number of sources, including specification of the problem, formulation of conceptual and computational models, estimation of input values, and calculation, interpretation, and documentation of the results. Of these, only uncertainties due to estimation of input values can be quantified with variance propagation techniques. Uncertainties that arise from mis-specification of the model can be assessed using decision trees and event trees based on elicitation of expert opinions. In some cases, using methods such as meta-analysis, model specification errors can be handled using simple variance propagation methods.

Nature of Models

Because the magnitude of chemical or microbial risks attributable to food can rarely be measured, such outcomes are estimated using models or projections from historical data. Exposure-effect models range from simple "rule-of-thumb" models to complex stochastic models. The reliability of these models is determined by the precision of the inputs and the accuracy with which the model captures the relevant biological, chemical, and physical processes. Uncertainty analysis can be used to assess how model predictions are impacted by model reliability and data precision.

Addressing Model Uncertainty

When there is uncertainty about the appropriate scenario or model, techniques can be used to assess the implication of alternate models on the predicted outcome. Methods such as probability trees, event trees, and fault trees can be used to portray the multiple events leading to the outcome of interest. An event tree starts with some initiating event and contains all the possible outcomes. The probability associated with each event may be represented by a probability distribution. The strengths of this approach include the visual portrayal of all the potential scenarios and the use of probability distributions as interpretations of relevant evidence.

Representing and Propagating Input Variance

Describing uncertainty in the risk involves quantification of the arithmetic mean value, the arithmetic or geometric standard deviation, and upper and lower quantile values of risk. Convenient tools for presenting such information are the probability density function or the cumulative distribution function for risk. However, the probability density function or cumulative density function of risk can often only be obtained when there are meaningful estimates of the probability distributions of the input variables used to estimate risk. There are five steps in an uncertainty analysis:

- identify inputs that could contribute to uncertainty in the predictions of a model;
- construct a probability density function to define the values that an input parameter can take;
- account for dependencies (correlations) among input parameters;
- propagate the uncertainties through the model to generate a probability density function of the outcome values; and,

— derive confidence limits and intervals from the probability density function of predicted values of the outcome variable.

The relationship between variance in model parameter inputs and the variance in the model predictions are estimated using variance propagation methods. Exact analytical, approximate analytical, and statistical simulation methods are available that can be used to propagate variance.

Uncertainty and Variability in Hazard Identification

The hazard identification step involves the determination that a health hazard is or may be associated with a biological, chemical, or physical agent present in foods. The step is generally based on screening methods and short and long-term cell or animal assays. Some examples and assay systems include quantitative structure-activity relationships, short-term bioassays, and animal bioassays. This step provides a dichotomous answer—that is, the factor is or is not thought to be a human health hazard. The uncertainty involves the correct classification of the agent (i.e., it is or is not a human health hazard) and performance of the assay in classification of the agent.

If the agent is evaluated in the assay multiple times, it is predicted to be either positive or negative with a certain degree of precision that is related to the performance of the assay. For example, one assay used to determine if a chemical is a mutagen is the Ames bacterial revertant assay. Uncertainty associated with the analysis of a chemical in this assay derives from knowing whether the assay is actually capable of predicting whether a positive response (or negative response) means that the chemical is capable (or incapable) of producing cancer in humans. The performance of the assay involves determining how the same chemical is characterised if analysed in this assay system at several different times and in different assay systems.

Three issues are considered potentially significant contributions to uncertainty and variability in hazard identification. First, is the misclassification of an agent—either identification of an agent as a hazard when it is not or the reverse. Second, is the issue of the reliability of the screening method both for appropriately identifying a hazard and the reliability of the assays to give the same result each time the assay is performed.

Third, is the issue of extrapolation because all screening methods are used to extrapolate the information provided by the test to predict human hazards. Epidemiological studies are used to predict the impact of exposures on human population in the future. As an example, in epidemiological studies, the extent of the extrapolation needed to predict health hazards for future human population is generally minimal; whereas, other assays have substantially greater need for extrapolation to produce predictions of potential adverse health effects for human population.

Uncertainty and Variability in Hazard Characterisation

Hazard characterisation is the process of defining the site, mechanism of action and at minimum the dose-response (proportion responding or severity of response) relationship. In this step, it is likely that a series of models may be developed. The models vary from purely mathematical representations to biologically-based representations. As a result, each has varying degrees of representation of the actual human disease process and as a result varying degrees of uncertainty.

Model uncertainty is likely to be an important issue in the hazard characterisation step. Mathematical dose-response relationships have the greatest uncertainty in actual representation of the biological processes. Despite the admitted large uncertainty, dose-response models are currently the most commonly used methods for predicting human health effects and have often proved useful in establishing policy. As interest in risk assessment has grown, the sophistication of the models, including the accuracy and completeness of their representation of the biological processes, has also grown.

An important issue of both variability and uncertainty that arises in hazard characterisation is in the variance in the dose-response at the dosage levels for the species studied. To increase power and the value of a negative study, typically large exposures are used in bioassays. These exposures are generally substantially greater than usual human exposures. That means that models including exposure response information gathered at high exposures may not be accurate at the low exposure levels of concern for human risk assessment. In addition, there is variance by animals in response at a given dose, despite the fact that most experimental animals are generally inbred and expected to be genetically identical. If outbred animals are used, the variability in the dose response relationship is expected to be larger, and if humans are exposed, the variance is also expected to be large.

Another issue of both uncertainty and variability that arises in hazard characterisation is the need to extrapolate between species. Approaches used for extrapolation between species include both uncertainty about the appropriate model for performing the extrapolation as well as variability in the parameters used for extrapolation.

Uncertainty and Variability in Exposure Assessment

Any model used to represent exposure should include several pieces of information:

- the level of an agent measured in a commodity or the levels measured in soil, plants, or animals that supply this commodity;
- the depletion/concentration ratio which defines changes in the level of an agent as a result of processing, preparation, and dilution;
- the frequency and magnitude of human intake of a commodity;
- the duration of contact or the fraction of a lifetime during which an individual is exposed to a commodity; and,
- the averaging time for the type of health effects under consideration to be clinically detectable.

These factors typically converge in the process of defining the distribution of population exposure. The population at risk for exposure refers to the population that consumes food containing the hazard. An exposure assessment is the key input to the assessment of dose, which reflects the amount of the agent delivered to the target organ or tissue, where the adverse effect can be induced

Defining exposure pathways is an important component of the exposure assessment. An exposure pathway is the course a biological, chemical, or physical agent takes from a known source to an exposed individual. In the case of agents in food, concentrations of chemicals and/or organisms can change between what is measured in soil, plants, animals and raw food and what is ingested by an individual.

In the case of chemicals, there can be some increases of contaminant concentration due to process (i.e. distillation), but more likely the storage, processing and preparation of the food product will result in a reduction of contaminant concentration. For organisms, there might be significant increases of microbe or contaminant concentration due to replication under favourable environmental conditions. Thus, significant uncertainties might be expected in the ratio of the concentration of a bacterial agent in food at the time of consumption to the concentration measured in raw foods or measured in animals, plants, or soil.

References

Covello VT, *Merkhofer MW Risk assessment methods: approaches for assessing health and environmental risks,* Plenum Press, New York, 1993.

Morgan MG, *Henrion M Uncertainty: A guide to dealing with uncertainty in quantitative risk and policy analysis,* Cambridge University Press, Cambridge, 1990.

Science and Judgment in Risk Assessment, *Committee on Risk Assessment of Hazardous Air Pollutants, Board on Environmental Studies and Toxicology,* National Research Council (NRC), National Academy of Sciences, National Academy Press, Washington DC, 1994.

Bibliography

Ackman, R.G., "Application of TLC to lipid separation: detection methods," In: *Analyses of Fats, Oils and Lipoproteins*, E.G. Perkins, American Oil Chemists' Society, Champaign, U.S.A. (ed.)., 1991, pp. 97-121.

Aitzetmuller, K., "The use of HPLC in modern fat analysis," *Fat Sci. Technol.*, 93, 501-510, 1991.

Amigo L., *et al.*, "Genetic polymorphism of ovine milk proteins: its influence on technological properties of milk" – a review, *International Daairy Journal*, 10: 135–149, 2000.

Bania J., *et al.*, "Application of polymerase chain reaction for detection of goats' milk adulteration by milk of cow", *Journal of Dairy Research*, 68: 333–336, 2001.

Caplen, R.H., *A practical approach to quality control*, Hutchinson Publishing Group, London, UK, 1982.

Cattaneo T.M.P., *et al.*, "Analysis of cow, goat and ewe milk mixtures by capillary zone electrophoresis (CZE): preliminary approach", *Milchwissenschaft, 51:* 616–619, 1996.

Covello VT, *Merkhofer MW Risk assessment methods: approaches for assessing health and environmental risks*, Plenum Press, New York, 1993.

Diaz, M.; Dunn, C. M.; McClements, D. J.; Decker, E. A., "Use of Caseinophosphopeptides as Natural Antioxidants in Oil-in-Water Emulsions", *J. Agric. Food Chem.*, 2003.

Guzey, D., McClements, D.J. & Weiss, J., "Adsorption kinetics of BSA at air-sugar solution interfaces as affected by sugar type and concentration", *Food Res. Int.*, 2003.

Hetherington, A.M. and Drobak, B.K., "Inositol-containing lipids in higher plants," *Prog. Lipid Res.*, 31, 53-63, 1992.

Hobbs, B.C. and Roberts, D., *Food Poisoning and Food Hygiene*, Edward Arnold, London, UK, 1987.

Hu M, McClements DJ, Decker EA, "Lipid oxidation in corn oil-in-water emulsions stabilized by casein, whey protein isolate, and soy protein isolate", *Journal of Agricultural and Food Chemistry*, 51, 1696-1700, 2003.

Khanna D.R. and Yadav P.R., *Biology of Fishes*, New Delhi, Discovery, x, 386 p., 2004.

Kramer, A. and Twigg, B.A., *Fundamentals of quality control for the food industry*, AVI Publishing Co., Westport, Conn. USA, 1962.

Laker, M.F. and Game, F.L. "Approaches to lipid and lipoprotein analysis," *Baillieres Clin. Endocrin. Metab.*, 4, 693-717, 1990.

Manuals of Food Quality Control, Vols 1-9, FAO Publications, Via delle Terme di Caracalla, 00100 Rome, Italy.

Morgan MG, *Henrion M Uncertainty: A guide to dealing with uncertainty in quantitative risk and policy analysis*, Cambridge University Press, Cambridge, 1990.

Patton,G.M. and Robins, S.J., "Extraction of phospholipids and analysis of phospholipid molecular species," *Methods Enzymol.*, 187, 195-215, 1990.

Saini V.P., Sharma L.L. and Ujjania N.C., *Dictionary of Aquatic Resources and Fisheries*, Udaipur, Agrotech, 2007, 648 p.

Science and Judgment in Risk Assessment, *Committee on Risk Assessment of Hazardous Air Pollutants, Board on Environmental Studies and Toxicology*, National Research Council (NRC), National Academy of Sciences, National Academy Press, Washington DC, 1994.

Shafi S.M., *Applied Fishery Science*, New Delhi, Atlantic Pub., 2003, 2 volumes, 601 p..

Shapton, D.A., and Shapton, N.F., *Safe Processing of Foods*, Butterworth-Heinemann, Oxford, UK, 1993.

Singh H.R. and Lakra W.S., (ed.)., *Coldwater Aquaculture and Fisheries*, Delhi, Narendra, 2000, 337 p.

Index

Acceptable Daily Intake (ADI) 243, 250
Active Oxygen Method (AOM) 110
American Oil Chemists Society (AOCS) 12
Association of Official Analytical Chemists (AOAC) 53

Babcock method 96

Chain traceability 220
Chemical spoilage 223
Critical Control Points (CCP) 217
Critical yield stress 175

Differential Scanning Calorimetry (DSC) 149
Differential Thermal Analysis (DTA) 156
Directorate of Fisheries (DOF) 216

Education act
Electromagnetic radiation 164
Endothermic process 152
Environmental Protection Agency (EPA) 2
Evaporation methods 138
Exothermic process 152

Fatty Acid Methyl Esters (FAMEs) 103
Fish industry trades 221
Fluorescence spectroscopy 165
Food and Drug Administration (FDA) 2
Food safety authorities 250

Generic interspecies scaling factors 249
Good Laboratory Practices (GLP) 246
Good Manufacturing Practices (GMP) 215, 219
Gravimetric instruments 154

Hazard Analysis Critical Control Point (HACCP) 256
Hazard identification 245, 262

Intra-unit variation 58
Iodine Value (IV) 104

Liquid extraction methods 95

Maximum Residue Limits (MRLs) 254
Multiplication rule 64

National Marine Fisheries Service (NMFS) 2
Non-polar solvents 92
Nuclear energy levels 161
Nuclear Magnetic Resonance (NMR) 106

Official analytical techniques 4
Organic solvent 92

Pathogenic bacteria 259

Qualitative risk assessment process 260
Quality assurance laboratory 7
Quality Index Method (QIM) 226

Rheological equipment 187
Rheological properties 179
Rheological thermal analysis techniques 156
Rodent-specific carcinogens 250

Sampling plan 52
Semi-continuous solvent extraction methods 93
Sinusoidal variation 162
Solid Fat Content (SFC) 110
Solvent extraction techniques 91
Specific heat capacities 158
Specific Spoilage Organisms (SSO) 225
Spectroscopy techniques 168
Standard Sanitation Operational Procedures (SSOP) 215
Statistical techniques 60

Thermal analytical techniques 153
Thermogravimetric techniques 148
Thin Layer Chromatography (TLC) 102
Thiobarbituric Acid Reactive Substances (TBARS) 109
Time-Temperature Indicators (TTI) 225
Total Viable Count (TVC) 225
Total Volatile Bases (TVB) 225
Toxicologically equivalent doses 249

United States Department of Agriculture (USDA) 2

Vibrational energy levels 164
Viscoelastic materials 182